CULTURE, PLACE, AND NATURE

STUDIES IN ANTHROPOLOGY AND ENVIRONMENT

K. Sivaramakrishnan, Series Editor

Centered in anthropology, the Culture, Place, and Nature series encompasses new interdisciplinary social science research on environmental issues, focusing on the intersection of culture, ecology, and politics in global, national, and local contexts. Contributors to the series view environmental knowledge and issues from the multiple and often conflicting perspectives of various cultural systems.

Forests Are Gold

Trees, People, and Environmental Rule in Vietnam

PAMELA D. McELWEE

UNIVERSITY OF WASHINGTON PRESS

Seattle and London

Publication of this book was supported by grants from the Association for Asian Studies First Book Subvention Program and the Rutgers University Faculty Research Council.

Printed and bound in the United States of Amercia
20 19 18 17 16 5 4 3 2 1

UNIVERSITY OF WASHINGTON PRESS
www.washington.edu/uwpress

Library of Congress Cataloging-in-Publication Data
Names: McElwee, Pamela D., author.
Title: Forests are gold : trees, people, and environmental rule in Vietnam / Pamela D. McElwee.
Description: Seattle : University of Washington Press, [2016] | Series: Culture, place, and nature | Includes bibliographical references and index.
Identifiers: LCCN 2015042773 | ISBN 9780295995472 (hardcover : alk. paper) | ISBN 9780295995489 (pbk. : alk. paper)
Subjects: LCSH: Forest policy—Vietnam—History. | Forest management—Vietnam—History. | Forest and community—Vietnam—History.
Classification: LCC SD657.V5 M33 2016 | DDC 333.7509597—dc23
LC record available at http://lccn.loc.gov/2015042773

The paper used in this publication is acid-free and meets the minimum requirements of American National Standard for Information Sciences—Permanence of Paper for Printed Library Materials, ANSI Z39.48–1984. ∞

This book is dedicated to the memory of Nguyễn Kiều Oanh and to my new daughter Riley Shea Duncan—although they never had a chance to meet, I know that they would have liked each other very much.

Contents

Foreword

THE FORESTS OF SOUTHEAST ASIA HAVE LONG PROVIDED A FERtile ground for developing a variety of fields of inquiry including environmental history, political ecology, cultural geography, and social studies of environmental sciences. Pamela McElwee joins an already crowded field of distinguished scholarship in one or more of these modes of studying Southeast Asia and its environment to inform wider scholarly debates. But with this book we now have a work that stands tall: dominating the canopy of the dense forest, many-hued in the flowers of research that may be found in it.

This wonderfully detailed and theoretically provocative study of forest management in Vietnam brings to fruition work that began in the late 1990s as McElwee visited and revisited the field sites, familiar and new, across Vietnam's highlands, with various questions. It is informed by new discovery that will literally reshape the environmental history of Vietnam as a field of study. To develop her argument, McElwee stays in the woods long enough to move through shifting paradigms in policy and public investment that remade forests from the source of raw natural resources into the fount of ecosystem services.

From French colonial rule to the most recent market-oriented socialism as Vietnam united, prospered, and entered the twenty-first century, it has become a powerful growing economy that has engaged in both ambitious conservation projects and rapid land conversion for a variety of development initiatives. In evaluating this transition, McElwee is equally attentive to the working of the state and other social institutions and to how forest management was experienced and occasionally shaped by the ordinary people caught up in its processes and outcomes. She is at once concerned with what government agents and social elites are doing, as well as deeply informed by the lives of farmers, forest dwellers, and ethnic minorities whose existence is rendered both visible and more precarious by the inclusion of their land and livelihoods in national pursuits.

Issues dear to the study of colonial and postcolonial forestry elsewhere in

Southeast and South Asia are visited and analyzed deftly, be it classification and silviculture, or the regulation of access for a variety of users and beneficiaries. McElwee finds that forestry was as much about forest management as it was about social control and exclusion and the appropriation of resources for various national and elite projects of development. But she also brings this synthetic work across the past century and into the present one through revisiting the same research areas. Along the way she is able to take on the topic of ecosystem services, for which forestry is increasingly being designated as a crucial sector in tropical and sub-tropical environments. Thus it is possible to see that forest management in Vietnam enables a process of environmental rule for resources and people, reordering landscape and society in the name of sovereign government and social assimilation of minorities.

The rise of environmental ideas and the spread of nature conservation programs across Asia occurred in a variety of political landscapes since the 1970s. In many instances as extant scholarship shows the programmatic and social outcomes reflected longer colonial legacies and more recent compulsions of national integration and economic development. How nature conservation interacted with pacification of frontiers, incorporation of minorities, and projects of national self-fashioning in Asia remains a topic of great interest.

Forests Are Gold provides a comprehensive study of these processes at work. McElwee draws well on the checkered and fast-changing political history of Vietnam, as colonialism, freedom struggles, war, revolutionary socialism, and globalization emerge as governing conditions through the twentieth century for determining the fate and use of forests. Long-term research in a rapidly transforming environment with forests renders this a masterful work that is brimming with insight for the next generation of social-ecological studies in Asia and beyond in a world that has been fundamentally altered by human action.

K. Sivaramakrishnan
Yale University
January 2016

Preface

WHEN I ARRIVED IN VIETNAM FOR THE FIRST TIME IN THE SUMMER of 1996, forests were on everyone's mind. The country was at a low point of tree cover, at around 25 percent of the total land area, which was bemoaned by officials as a mere half of the forest estate at the twilight of colonial rule. Vietnamese newspapers regularly published stories about the nefarious deeds of illegal loggers, and several extensive floods focused further attention on the presumed link to deforestation. The official concern over forest loss was so great that the government mobilized nearly two billion US dollars in domestic and international funding for a massive campaign to restore tree cover to 43 percent of the country's land area, which was the amount of forest posited to exist in the mid-1940s. Just twelve years after the campaign to reforest five million hectares of forests began in 1998, the government claimed they had met their goal.

There have been many postmortems on how (as well as if) Vietnam achieved this outcome and what other nations can learn from these lessons (de Jong 2009; Lambin and Meyfroidt 2010). Vietnam has been lauded as an example of a "forest transition," whereby states move from net deforestation to net afforestation as they develop (Mather 2007). Forest transition theory draws heavily on the "environmental Kuznets curve" postulate, which suggests an inverse U-shaped curve relating environmental degradation to income; poor countries experience rising degradation and deforestation during early developmental processes, but such patterns are replaced by environmental cleanup and reforestation once countries reach middle-income status (Bhattarai and Hammig 2001). The assumption from these theories is that Vietnam has undergone a forest transition because it has strengthened its market economy since the 1980s, and thus can be a green success story for other nations to emulate, if only they would follow a similar development path.

But as I undertook fieldwork throughout this period in Vietnam, two things struck me as not only incongruous, but incorrect, about this explanation. First, while it was true that the country was planting trees on a monumental scale, this coincided with continuing high deforestation rates in other areas (primarily biodiverse natural forests). At one point in the late 1990s, Vietnam had both the tropical world's second highest deforestation rate after Nigeria, as well as the third highest afforestation rate after French Polynesia and Rwanda. That a country could simultaneously plant and deforest so many trees was not addressed anywhere in forest transition theory, and it stood out as very strange.

Second, I had the opportunity to watch the reforestation program in action during fieldwork in 2000–2001 as funding to encourage households to plant trees trickled into the area of north central Vietnam where I was working. What I saw on the ground was not an environmentalists' dream of expanding green forests, but rather a nightmare of overreliance on introduced fast-growing but low-value trees that displaced native flora and fauna. The afforestation, mainly by monocropped exotic Australian eucalypts and acacias, could not hide the continuing degradation of natural forests in a nearby nature reserve. Social changes accompanied the spread of these plantations as well, and from the local point of view, the forest transition was a process that involved struggle and contestation, not a linear pathway from fewer trees to more trees. Tenure over newly planted trees was highly contentious, and those with power and access were getting benefits from afforestation while the poor often did not. Gender relations too were unequal, with women losing rights to land as men claimed it for new forest plantations, and the promised poverty reduction benefits from the tree-planting program failed to materialize for many.

Eventually I came to a recognition that these outcomes were linked, and not unexpected. Afforestation policy was not really aimed at "improving" the landscapes of Vietnam for biodiversity or conservation's sake. Rather, it was a policy with other social and economic goals in mind: the shifting of responsibility for large areas of land from previous state-managed cooperatives to households, the accessing of international development dollars to stave off unemployment in a declining forest sector, and the expansion of low-quality wood supplies for local sawmills and pulp factories. Because the afforestation policy was not in fact aimed at, and did not actually address, the underlying drivers of environmental change and pressures on forests, it was no surprise that reforestation and deforestation could coexist, or that the outcomes of afforestation practice could be so ecologically and socially questionable.

Many commentators have assumed that environmental policy is primarily aimed at improving the natural world, such as reducing forest degradation and conserving biodiversity. This book explores why that assumption is often wrong. As I will show, environmental policy is at times aimed not at nature, but at people, and failing to acknowledge this fact has resulted in numerous unintended, not to mention some intentional, consequences. Vietnam provides a prime example of this problem in action. While there has been rising concern about processes of environmental change in Vietnam, such as "deforestation," and the development of many interventions and policies deemed as "environmental" to combat these changes, this has been accompanied by the dispossession of forest land, particularly for poor and marginalized groups, the rise of wealthy mafia-style timber smugglers, and the continued loss of species and biodiversity, even from so-called "protected" areas. These effects are not coincidences. They are outcomes of a situation I term "environmental rule," whereby states, organizations, and individuals use environmental explanations to justify policy interventions in other social areas, such as populations, markets, settlements, or cultural identities. This book outlines what environmental rule is, how it develops, and how it can be analyzed, using the case study of forests in Vietnam.

Although I have been trained as an anthropologist as well as an environmental scientist, this book is not an traditional ethnography per se. I make use of multiple sources, including extensive work in archives in Vietnam and France, and numerous interviews and interactions with government officials, international conservationists, and everyday citizens to contextualize how forests have been subjected to various forms of environmental rule. I did most of my fieldwork in 2000–2001 in Hà Tĩnh province, approximately 300 kilometers south of the national capital of Hanoi, along the narrow central coast of Vietnam, therefore much of the book refers to this area, once known as Annam under French colonialism, and which was part of the Democratic Republic of Vietnam (DRV, or North Vietnam) after 1954. Later I was able to extend fieldwork to the provinces of Quảng Trị, Thừa Thiên Huế, Quảng Nam, and Đắk Lắk in 2005–06 and Lâm Đồng Province in 2011–13, which were part of the former South Vietnam (the Republic of Vietnam, RVN) from 1954 to 1975. However, the book by necessity focuses much more on the experiences of environmental rule in the DRV than the RVN.

My fieldwork in 2000–2001 focused on rural areas of Hà Tĩnh, and was aimed at understanding state and local management of forests around a protected area (the Kẻ Gỗ Nature Reserve, or KGNR). Hà Tĩnh was historically important in several ways: it was the province in Annam with the most forest reserves demarcated under French colonialism, and, in a not-unrelated development, the site of well known protests of 1930, the Nghệ

Tĩnh Soviets, that gave rise to the Indochinese Communist Party as a serious political movement. Hà Tĩnh was also the site of the discovery of new mammals previously unknown to science in 1992, such as the saola (*Pseudoryx nghetinhensis*), which established the area's reputation as a site in need of biodiversity conservation.

Two major nature reserves, the Vũ Quang National Park and the KGNR, were demarcated within Hà Tĩnh to protect the newly discovered biodiversity in the 1990s. The KGNR was described at the time of founding as having one of the "largest remaining blocks of broadleaf evergreen forest in the level lowlands of central Vietnam," and according to biologists from Birdlife International, 46 species of mammal, 270 species of bird, and 562 species of plant could be found there (Lê Trọng Trải et al. 2001, vii). Despite the grand-sounding description, however, in fact the KGNR was heavily disturbed forest, as it had been the site of past logging by numerous state-owned timber companies for decades, until the area was declared a Watershed Protection Forest in 1990. Limited logging still occurred up to 1996 when the area was converted to a nature reserve, one of over one hundred new protected areas that were proclaimed in Vietnam in the late twentieth century.

The KGNR was not alone among the many protected areas that were created out of the ashes of over-logged former timber reserves, and most estimates agree that Vietnam has very few remaining forests that could be classified as primary or undisturbed (FAO 2010). The KGNR was also typical of many protected areas in Vietnam in that thousands of people lived near the park boundaries and exerted pressure on the reserves' resources, which were now to be strictly protected behind fences and rangers. To understand forest use around the reserve, and why the new park was so contested, I worked in five main villages along the buffer zone, interviewing households as well as local officials and rangers. I also implemented a standardized quantitative survey on forest and land use with 104 participating households (the main results of these surveys can be found in McElwee 2008 and 2010).

In 2004–5, I returned to Vietnam for additional fieldwork on migration, conservation and forest management around several protected areas in central Vietnam. These studies were aimed at assessing the livelihood impacts of local ethnic minority populations on the nature reserves, as well as the pressures of Vietnamese migration into these sites. I also had the opportunity in 2005–6 to advise the Institute for Ethnic Minority Affairs, a government research institute, on how to professionalize their research practices, and, as a result, I was able to travel to a number of other sites in Vietnam to carry out research with them on major issues like swidden agriculture and sedentarization policies for ethnic minorities.

In the summers of 2008 and 2009 I returned to Vietnam and interviewed stakeholders in Hanoi regarding policymaking on forest and biodiversity issues, including a number of national figures, such as heads of departments at the Ministry of Natural Resources and Environment (MONRE) and the Ministry of Agriculture and Rural Development (MARD), the former head of the national forest ranger service (*Kiểm lâm*), congresspeople in the National Assembly (Vietnam's legislative chamber), and heads of international and local conservation NGOs and donor-funded conservation projects. My research assistant and I carried out over fifty policymaker interviews in total. Finally, since the fall of 2011, I have been working on a new research project on payments for environmental services (PES) and forest carbon sequestration policy (known globally as Reduced Emissions from Deforestation and Degradation or REDD), together with a team of researchers from the Center for Natural Resources and Environmental Studies and the NGOs PanNature in Hanoi and Tropenbos International Vietnam in Huế city, and we have surveyed over four hundred households in Lâm Đồng, Sơn La, Thừa Thiên Huế, Kon Tum, Kiên Giang, and Điện Biên provinces on PES and REDD policy.

In addition to this local fieldwork, I have made use of information collected at the Centre des Archives d'Outre-Mer (CAOM) in Aix-en-Provence, France in 2003; the National Archives of Vietnam Number 1 (NAV1) in Hanoi in 2008 and 2009; the National Archives of Vietnam Number 2 (NAV2) in Ho Chi Minh City in 2007; the National Archives of Vietnam Number 3 (NAV3) in Hanoi in 2009; and the National Archives of Vietnam Number 4 (NAV4) in Dalat in 2009. As many researchers have discovered, using the documents in the various branches of NAV can be challenging, especially in NAV3 where many files are not in indexes or else are deemed too sensitive for foreign eyes. Luckily forests and conservation did not fall into this latter category (an interesting fact in and of itself), and some of the documents on forest policy referenced in chapters 2 and 3 have not to my knowledge been seen or used before in the English language scholarship on Vietnam. (All translations from documents in Vietnamese, French, and German used in this book are my own, unless noted otherwise.)

Additional publications were consulted at the National Library of Vietnam in Hanoi, the Science and Technology Library of Ho Chi Minh City, the NGO Resource Library in Hanoi, the Social Science Information Library in Hanoi, the United Nations Development Programme Library in Hanoi, and the small libraries of grey literature held by the Food and Agriculture Organization, the World Wildlife Fund, and the International Union for the Conservation of Nature, all in Hanoi. Supplemental documents were obtained from the very substantial Vietnamese language holdings of Sterling Memorial Library at Yale, the Kroch Library at Cornell, the library at the School of Oriental and African Studies in London, the Library of Congress, and the Arizona State University library.

All of these sources have provided the evidence for my analysis of environmental rule. My goal in this book is not to identify an exact typology of environmental rule. Rather, my goal is to illuminate what environmental rule can look like and how it might be analyzed, whether by examining the case study of forests in Vietnam here, or, as I hope, for my arguments to be taken up by others to explain causes and consequences of environmental change and social policies in other places where environmental rule can be found.

Acknowledgments

WHEN ONE TAKES OVER TEN YEARS TO FINALLY GETTING AROUND to completing a book, an excessive number of debts pile up. I have many thanks to multiple funders. Grants for field research from 1999 to 2001 were provided by the Social Science Research Council; a Wenner Gren Foundation Small Grant; and a National Science Foundation Cultural Anthropology grant no. 0108992, "The Effects of Ethnicity and Migration on Forest Use and Conservation in Central Vietnam." Additional funding was provided by the Yale Center for Biospheric Studies Hutchinson Fellowship, the Yale Center for International and Area Studies Henry Rice Hart Fellowship, a Charles Kao Fund grant from the Yale Council on East Asian Studies, the Yale Program in Agrarian Studies, and the Yale Council on Southeast Asian Studies. I was able to write up fieldwork findings thanks to the generosity of a Yale University Writing Fellowship, a Teresa Heinz Scholars for Environmental Research Fellowship, and a Switzer Foundation fellowship.

In 2004–5, I returned to Vietnam as co-principal investigator with Chris Duncan for a project on "Environmental Consequences of State-Sponsored Rural-Rural Migration in Southeast Asia: A Comparison of Transmigration and Resettlement in Indonesia and Vietnam," generously funded by the John D. and Catherine T. MacArthur Foundation's Program on Global Security and Sustainability Research and Writing Grant. From 2006 through 2011, I received several internal grants from Arizona State, including a faculty development grant from the Center for Asian Studies, which enabled me to undertake archival research. A Global Engagement Seed Grant from the Office of the Vice President also funded additional interview work. In the summers of 2008 and 2009, I received funding from a John D. and Catherine T. MacArthur foundation grant to the "Advancing Conservation in a Social Context" project.

Since 2011, a National Science Foundation (NSF) Geography and Regional Science Division grant no. 1061862 has generously provided support for our project "Downscaling REDD policies in developing countries:

Assessing the impact of carbon payments on household decision-making and vulnerability to climate change in Vietnam." My position at Rutgers has also been supported by the New Jersey Agricultural Experiment Station, and specifically Hatch funding from the National Institute for Food and Agriculture. My Vietnamese collaborators on this project were additionally funded by the Economy and Environment Program for Southeast Asia for fieldwork in 2011, and since 2012, they have also been supported by a US Agency for International Development Partnerships for Enhanced Engagement in Research grant titled "Research and Capacity Building on REDD+, Livelihoods, and Vulnerability in Vietnam: Developing Tools for Social Analysis and Development Planning."

Anyone who has worked in Vietnam knows the importance of local connections and sponsorships, which will make or break research success. I have been enormously lucky in getting access to an extraordinary range of research sites, for which I have numerous people to thank. First, Dr. Võ Quý has been Vietnam's preeminent conservationist for decades, and I have been honored to know and work with him. I hope I have done justification to his home province in the pages herein. For my work in Hà Tĩnh, Võ Thanh Giang, Dr. Quý's son, was my key contact: Giang went to numerous meetings to make sure I got and kept my permission to work there at a time when almost no other foreign researchers were allowed out in rural areas. Dr. Trương Quang Học, former head of the Center for Natural Resources and Environmental Studies at Vietnam National University, deserves my hearty thanks for agreeing to help and sponsor my research when it seemed like no one else was interested. Dr. Hoàng Văn Thắng, the current head, has taken the baton and continued the strong support.

I have also been tremendously fortunate to have a tight-knit group of collaborators in the past few years. Dr. Nghiêm Phương Tuyến, Dr. Lê Thị Vân Huệ and Vũ Thị Diệu Hương have worked closely with me on a number of projects since 2006, some on forests, some on climate change, and the results of each have been made better by their diligence, kindness, and above all, senses of humor. Hương also served as my key research assistant during 2008–9 in interviewing biodiversity policy makers. Trần Hữu Nghị of Tropenbos International Vietnam has also been an invaluable collaborator, and his support was crucial for fieldwork in 2005, as well as on our current carbon forestry project. All these friends have made work in Vietnam the past few years not just interesting, but enjoyable.

My fieldwork in Hà Tĩnh in 2000–2001 was made possible with the help of a wide number of people: Đặng Anh Tuấn, Trương Thanh Huyền, Nguyễn

Thái Bình, and Lê Duy Hưng served as research assistants at different points in the project. A number of people in Cẩm Xuyên district also made my fieldwork possible: first of all, Trần Văn Sinh and Trần Thị Kim Liên of the Non-Timber Forest Product project field office and field officers Mr. Trần Đình Duy, Mr. Danh Viết Vị, and Mr. Hà Huy. During fieldwork in 2004–05 on ethnic minority resource use in Quảng Trị, Đào Nguyên Sinh of the Forest Inventory and Planning Sub-Institute in Huế city and (now Dr.) Vũ Thị Hồng Anh of Syracuse University were helpful research assistants. Since 2011, administrative assistance for our NSF-funded project has been provided by Dr. Đào Minh Trường, Lê Trọng Toán, and Hà Thị Thu Huế of CRES and Hà Thị Tú Anh of Tropenbos International Vietnam. Phạm Việt Hùng, Nguyễn Minh Hà, Đặng Tú Loan, and Vũ Thị Minh Hoa of CRES and Nguyễn Việt Dũng, Trịnh Lê Nguyên, Nguyễn Xuân Lãm, and Nguyễn Hải Vân of PanNature have also been instrumental in the project. Assistance was also provided in our field site of Lâm Đồng by Mr. Đỗ Mạnh Hùng of Bi Đúp Núi Bà National Park.

Additional thanks go to individuals in Vietnam who have helped in other ways over the years, through discussions, documents, and patiently answering my questions: Võ Trị Chung and Phùng Tửu Bôi of the Forest Inventory and Planning Institute; Dr. Lê Ngọc Thắng, Dr. Lê Hải Đường, and Dr. Phan Văn Hưng of the Institute for Ethnic Minority Affairs; Hà Hoa Lý of the Ho Chi Minh Political Academy; Koos Neefjes of the United Nations Development Program; Dr. Vương Xuân Tình of the Institute for Anthropology; and Dr. Lê Trần Chấn of the Biogeography Division of the Institute of Geography in Hanoi, who very generously identified my plant specimens.

In the eternal struggle to speak Vietnamese properly, I have many people to thank (and some to blame). I first encountered Vietnamese with Uncle Long Nguyễn, a New Haven resident who generously tutored me in my first year at Yale, and a series of excellent teachers in Hanoi, including Thầy Bình, Cô Hương, Cô Đài, Cô Quyên, and Cô Oanh. I particularly want to thank Cô Nguyễn Kiều Oanh and her entire extended family, especially Hồ Quỳnh Giang, for being my surrogate family. Cô Oanh and Giang were more like my mother and my sister than my teacher and her daughter-in-law whenever I was in Vietnam. Cô Oanh sadly passed away in fall of 2014 just as this book was heading toward publication, so she was not able to see how her wayward pupil had finally improved, but I hope I have done justice to her considerable efforts to make me as Vietnamese as possible. My life in Vietnam was also always made better by several families with whom I lived in Hanoi, particularly the family of Bác Cao Xuân Chử in 1996–97, and especially my landlady (and friend) throughout much of my time in Vietnam from 1999–2012, who always made a room in her house open to me when I was in Hanoi, Chị Đỗ Thị Quỳnh.

As this work has gone through various iterations, I want to thank the many people who read, listened, commented, and often challenged me on this material. My early mentors Michael Dove, Eric Worby, K. Sivaramakrishnan, James C. Scott, and Helen Siu began pushing me down this path. Others who read through my work carefully and sent me in new directions in graduate school include Nancy Peluso, Hal Conklin, and Arun Agrawal. Various ideas and iterations of the chapters in this book that I gave at conferences and talks over the years have been commented on by James Rush, George Thomas, Peter Zinoman, Peter Brosius, Paige West, Erik Harms, Shivi Sivaramakrishnan, Philip Taylor, Hy Van Luong, Oscar Salemink, Arun Agrawal, Ashwini Chhatre, Raj Puri, Andrew Mathews, Ben Kerkvliet, and Ben Orlove, and I hope I have included their many helpful suggestions satisfactorily. I also greatly appreciate the comments from the students and faculty who attended talks I gave at the Cornell Southeast Asian Studies program (April 2004), the Berkeley Southeast Asian Studies Program (November 2006), the Columbia University Workshop on Politics, Society, Environment, and Development (September 2006), the Uppsala University workshop on Climate Change, Environment and Society in Southeast Asia (August 2010), the Vietnam Update at the Australian National University (November 2011), the Yale Program in Agrarian Studies (February 2012), the Boston University Global Development Seminar (April 2012), and the Rutgers University Geography department (April 2013). Several people very generously read some or all of this final manuscript and provided very useful comments which strengthened my arguments considerably: Karen O'Neill, Tom Rudel, Bernhard Huber, and Mitch Aso. I also thank the two reviewers whose thoughts were immensely useful in shaping the final book, along with the editorial guidance of Lorri Hagman and Shivi Sivaramakrishan. Vũ Thị Hồng Anh and Trần Hữu Nghị did final checks of the Vietnamese spellings in the manuscript. Subventions from the Rutgers University Faculty Research Council and the Association for Asian Studies First Book Subvention Program helped make publication possible.

General thanks are also due to a number other people who have helped along the way with additional ideas, suggestions, sources, and sometimes just a distraction from writing: Mila Rosenthal, Jenny Sowerwine, Jane McLennan, Thomas Sikor, Mark Poffenberger, Mike Arnold, Nina Bhatt, Vũ Thị Hồng Anh, Huỳnh Thu Ba, Dương Bích Hạnh, Hoàng Cầm, Tô Xuân

Phúc, Patrick Meyfroidt, and Kathleen Abplanalp. Charles Keith, who gave assistance in navigating the French archives in 2003, was deeply appreciated by this non-historian.

At Yale, my path was made smoother by Kay Mansfield at Agrarian Studies, Kris Mooseker at Southeast Asian Studies, Elisabeth Barsa at the School of Forestry and Environmental Studies, and Rich Richie at Sterling Library. At Arizona State University, Gisela Grant provided useful administrative assistance. At Rutgers, the administrative support of Kristen Goodrich, Justine DiBlasio, and Wendy Stellatella has made work and travel to Vietnam more manageable. I also thank William Hallman, the chair of the Department of Human Ecology, for a semester's leave in fall 2011 to enable me to spend time in Vietnam. My colleagues in the department of Human Ecology at Rutgers, particularly Karen O'Neill and Tom Rudel, have also been exceedingly helpful in reading through various materials I needed feedback on. Dean Bob Goodman of the School of Environmental and Biological Studies has also been highly supportive since my arrival at his school in 2011.

My parents, Carl and Marge McElwee, deserve thanks for being my mail and money managers while I was away from the United States on many occasions, and for being a supportive presence generally. My sister Heather came to visit me during my last two weeks of fieldwork in 2001, and she was an excellent bag-carrier and travel companion as we carried out 200 pounds of books and surveys on my exit. And finally, my longtime partner in Southeast Asian crime, my husband Chris Duncan, has been my best friend and strongest supporter since we met in graduate school, including residing with me in Vietnam in 2004–5. He has read through much of this work, but mostly he has listened to me complain about various obstacles in fieldwork and writing for over fifteen years, and, for that, he deserves an enormous gold medal. What he will get instead is the completion of this book, along with our first child, Riley Shea, who arrived just as these pages were going off to press, which I hope will suffice as a down payment on my many debts of gratitude to him.

Vietnamese Terminology

A NOTE ON VIETNAMESE SPELLING

Because Vietnamese requires the use of diacritics to distinguish between words with similar spellings, I use these diacritics for most words and place names throughout the book, with the exception of words that are familiar to English readers: for example, Vietnam rather than Việt Nam, Hanoi rather than Hà Nội, and Ho Chi Minh, rather than Hồ Chí Minh. Where Vietnamese proper names do not have diacritics (such as some sources in the bibliography), it is because the authors themselves choose not to use them.

KEY VIETNAMESE TERMS USED THROUGHOUT THE TEXT

đất trống empty lands, waste lands

Định Canh Định Cư The Fixed Cultivation and Sedentarization Program

đổi mới Renovation, an open door economic and political policy

đồi trọc Bare hills

Khai hoang "Clear the Wilderness" program

Kiểm lâm Forest Protection Department, or forest ranger service

Kinh Ethnic Vietnamese

lâm tặc Illegal logger

lâm trường State Forest Enterprise, or state-owned logging company

rẫy Swidden agriculture

rừng Forest

Rừng là vàng "Forests are gold," a quote from Ho Chi Minh

sào Local land measurement unit, equivalent to 360m^2 in northern Vietnam, 500m^2 in central Vietnam, and 1000m^2 in southern Vietnam. In this text, 500m^2 is the standard.

Việt Minh League for the Independence of Vietnam, led by Ho Chi Minh

Abbreviations

ARBCP	Asia Regional Biodiversity Conservation Program
ANT	Actor-Network Theory
CAOM	Centre des Archives d'Outre-Mer, France
CHER	Cultural, Historical, Environmental and Landscape Reserve
DRV	Democratic Republic of Vietnam (also known as North Vietnam)
EVN	Electricity of Vietnam
FCPF	Forest Carbon Partnership Facility
FCSP	Fixed Cultivation and Sedentarization Program
FIPI	Forest Inventory and Planning Institute
5MHRP	Five Million Hectare Reforestation Program
FPDF	Forest Protection and Development Funds
FULRO	*Front Unifié de Lutte des Races Oprimées* (Unified Front for the Liberation of the Oppressed Races)
GGI	Government General of Indochina
ICP	Indochinese Communist Party
ICRAF	World Agroforestry Center
KGNR	Kẻ Gỗ Nature Reserve
KL	*Kiểm lâm*: Forest Protection Department
MARD	Ministry of Agriculture and Rural Development
MEA	Millennium Ecosystem Assessment
MOF	Ministry of Forestry
MOIT	Ministry of Industry and Trade
MONRE	Ministry of Natural Resources and Environment
NAV	National Archives of Vietnam
NTFPs	Non-Timber Forest Products

NXB	*Nhà Xuất Bản*: publishing house
PAM	*Programme Alimentaire Mondial*: World Food Program
PES	Payments for Environmental Services
REDD	Reduced Emissions from Deforestation and Degradation
RSA	*Résident Supérieur* of Annam
RST	*Résident Supérieur* of Tonkin
RUPES	Rewarding Upland Poor for Environmental Services project
RVN	Republic of Vietnam (also known as South Vietnam)
SFE	State Forest Enterprise
SIDA	Swedish International Development Agency
SRV	Socialist Republic of Vietnam
STS	Science and Technology Studies
SWAT	Soil and Water Assessment Tool
UN-REDD	United Nations' Reduced Emissions from Degradation and Deforestation Program
USAID	United States Agency for International Development
VND	Vietnamese *đồng*, the currency unit of Vietnam
VNFF	Vietnam Forest Protection and Development Fund
VNFOREST	General Directorate of Forestry
WWF	World Wildlife Fund

Forests Are Gold

Introduction

Seeing the Trees and People for the Forests

HO CHI MINH, VIETNAM'S MOST WELL-KNOWN REVOLUTIONARY leader, once famously remarked, "Forests are gold; if we know to protect and develop them well, they will be very precious."[1] For many years, forest rangers and government officials told me that this phrase was uttered by Ho in 1962, during the dedication of Cúc Phương National Park, the independent state of North Vietnam's first designated protected area. This expression of environmentally conscious engagement was striking given the timing: in the early 1960s conflict with South Vietnam was escalating into what would become the US–Vietnam War. The fact that the president of North Vietnam would take time off from revolutionary struggles to personally attend the park's opening ceremony, I was told on multiple occasions, was a testament to the country's emergent conservationist sensibilities. Ho's shortened catchphrase, "forests are gold" (*rừng là vàng*), subsequently became a slogan after his death, which was repeated in government campaigns to plant trees and protect biodiversity (see figure I.01).

I soon discovered that this version of the story was not quite right. There is no record of Ho's speech at Cúc Phương in 1962, or of his even making that trip. But he did say "forests are gold" at an event in 1963 that had nothing to do with biodiversity or conservation. On August 31, President Ho met with two-hundred delegates to the Mountainous Areas Party Education Conference in Hanoi. The conference aimed to review propaganda work and assess what was needed to move solidly towards socialism, and Ho exhorted cadres to improve their efforts to spread party ideology into the mountainous hinterlands, especially among the ethnic minorities that made up much of the population. His talk covered a wide range of issues, from how to introduce theoretical and political concepts to the masses to how to expand collectivized agriculture. By the middle of the speech, Ho also insisted that environmental campaigns were part of the work of making new citizen-subjects. He urged the assembled delegates,

FIGURE I.01 Ho Chi Minh's phrase "Forests are gold; if we know to protect and develop them well, they will be very precious" at the entrance to a protected watershed forest in Lâm Đồng Province, 2014. Photo by author.

> All of you need to pay urgent attention to forest protection. If the situation continues where our ethnic compatriots destroy a little, then the state agricultural farms destroy a little, then industrial farms destroy a little, even if a survey team that is inspecting local geology also destroys a little, then it is damaging. Destroying the forest is easy, but bringing about forests again will take us decades. Large-scale deforestation in this way will cause great impacts: on climate, on production, and on life. We often say 'Gold forests, silver seas.'[2] Forests are gold; if we know to protect and develop them well, they will be very precious.[3]

This vision of a strong nation, encompassing marginalized ethnic minorities in the remote borderlands, united in protecting and replanting forests for climate regulation and agricultural production, contrasts with the image that contemporary environmentalists provide of President Ho laying the foundations for wildlife and biodiversity preservation for conservation. Yet the speech encapsulates much that is noteworthy about forest policy in Vietnam, particularly the focus on the confluence of nature, the state, and citizens. The overall gist of Ho's message is clear: forests are about more than trees. They are about the management of people as subjects and nature as an object.

Environmental policy is often presented as a tool to shape "things," such as plants, soils, or water, rather than people. What this study of forest management in Vietnam over the last century instead reveals is that environmental interventions have never been exclusively about "nature" or ecology, but rather about people and society, as Ho Chi Minh suggested. Although labeled as "environmental," many policies and actions directed at forests were in fact about the supply of wood for a war-torn nation, the movement and control of people in sensitive remote areas, or the prolonging of state employment in a declining forest sector. In other words, forest policy in Vietnam has rarely been about ecological management or conservation for nature's sake, but about seeing and managing people, a strategy I term "environmental rule."

Environmental rule occurs when states, organizations, or individuals use environmental or ecological reasons as justification for what is really a concern with social planning, and thereby intervene in such disparate areas as land ownership, population settlement, labor availability, or markets. Imposing a vision on landscapes has always been a role of the state (Scott 1998; Sivaramakrishnan 1999; Peluso and Vandergeest 2001). What is unique about environmental rule is that while the justification for intervention is to "improve" or "protect" the environment itself, in reality, underlying improvements to people or society are envisioned. For example, a policy on watersheds may really be about resettling ethnic minorities perceived to be opposed to the state; a policy on restricting timber sales may be about controlling revenues for a forestry agency rather than stopping illegal logging. One practical example can be seen in the treatment of the practice of swidden (or shifting) agriculture, a long-standing system among many highland dwellers in tropical Asia to rotate or interplant agricultural crops with forests (Dove 1983), but which has been opposed by all political authorities in successive regimes in Vietnam. Attempts to control or prevent swidden agriculture have always relied on ecological explanations, such as avoiding soil erosion on slopes or loss of biodiversity in cultivated landscapes, but in nearly every case, the ecological "science" underlying these justifications has been sparse, while the social pressures to assimilate ethnic minorities to ethnic majority customs, resettle them out of economically valuable forests, or simply extract more tax revenue from them, have been the clearest concerns driving so-called "environmental" policy.

Environmental rule provides a useful lens for understanding policy interventions and outcomes that at first glance might appear incongruous or inconsistent. The concept offers a clearer explanation for the interventions directed at nature, which have not been confined to linear patterns of capitalism, socialism, or neoliberalism, as others have asserted (Prudham 2004; Humphreys 2008; Peet et al. 2011). Rather, policies for the environment often emerge out of unexpected relational interactions between people, ideas, and objects—what have been called "networks of rule" (Miller and Rose 2008). By comparing and contrasting these myriad processes and networks we can see how different projects of environmental rule are formed, how they operate, and how they transform action on the ground (or fail in these attempts, as the case may be). Environmental rule is not static, but transforms over time. Forest policies under the French colonial regime in Indochina were characterized by top-down state power and undemocratic coercion in the demarcation of forest reserves and restrictions on woodland use by locals. Yet by the end of the twentieth century, discussions of free markets and local participation, with an emphasis on voluntary and individual efforts for forest carbon conservation, dominated policy. In both cases, however, what looked on the surface like interventions directed at forests were in reality directed at changing the location, conduct, and even the identities of people themselves.

The shifts in focus of environmental rule over time, and of the people and institutions that enforce or engage in ruling and those subject to it, also highlights the fact that environmental rule is always co-produced in the sense that both natural and social formations are created and mutually reinforce each other, and that objects of rule (whether people or trees) often have their own ideas which differ from those of the rulers.[4] Environmental rule does not get implemented in a vacuum—it is influenced by global knowledge networks that circulate shifting ideas, concepts and classifications about "nature" or the "environment," the resistance of human subjects to doing what they are told to do, and even by the physical properties of the environment itself. Co-production implies that we cannot look solely for power and force in our discussions of environmental rule; we must seek out the ways representations, identities, knowledge, and culture meld together to subtly direct human action towards some goals and not others, and the important effects that physical constraints, such as tree growth rates, water supplies, or soil chemical content might have on nature-society interactions.

To show how environmental rule arises and evolves, this book assesses forest policy and management in Vietnam during the long twentieth century, from roughly 1884, the beginning of formal French colonial rule, to the present day, and draws on several years of anthropological fieldwork in rural areas of Vietnam (see map I.01), extensive interviews with numerous state and local officials, NGOs, and donors, and archival work in Vietnam and

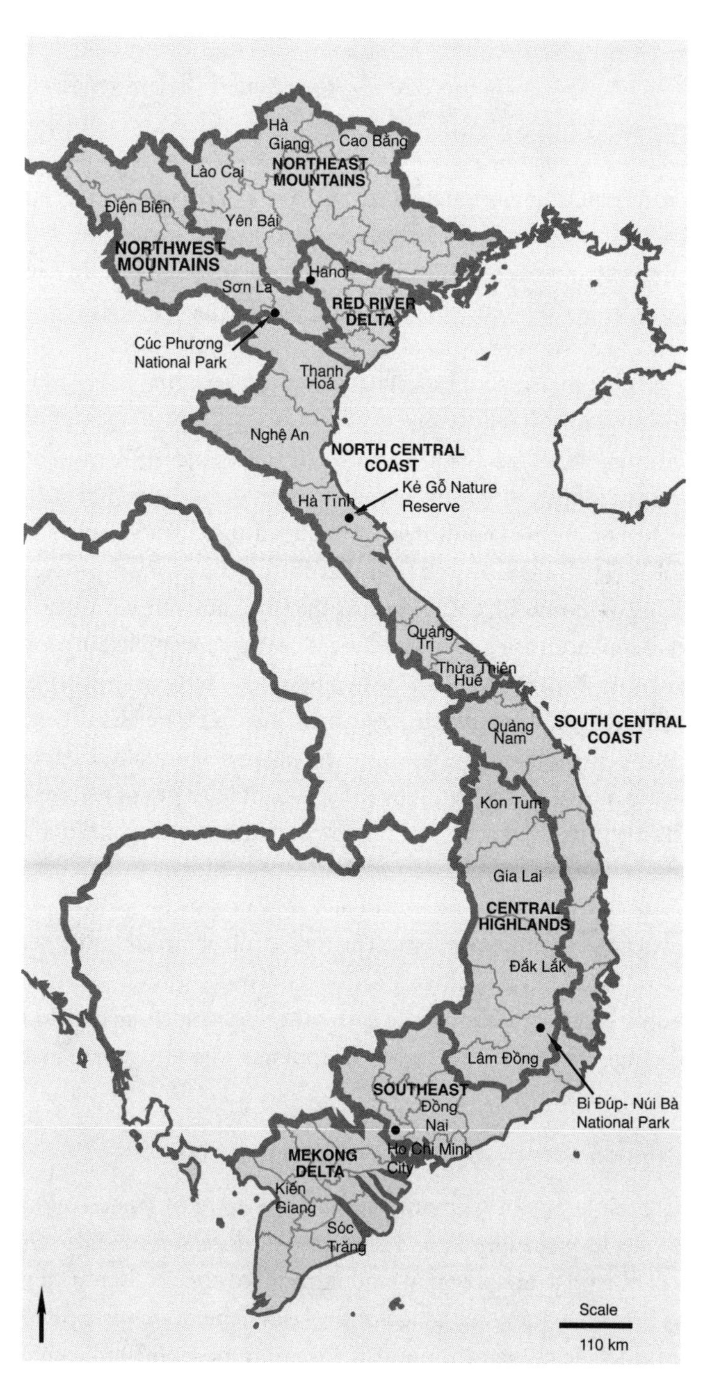

MAP I.01 Regions and provinces of Vietnam showing locations of fieldwork mentioned in text. Base map from Vietnam location map by Uwe Dedering on Wikimedia Commons. Map redrawn by author.

France. This historical-ethnographic approach contrasts with other histories and policy studies, which often take for granted the existence of a clear problem like "deforestation," which is then explained in terms of a theoretical approach like population growth (Kummer 1992). Instead, thinking of forest policy as a form of environmental rule asks how particular issues come to be posed as problems in need of solutions (that is, as justifications for "rule") in the first place. Epistemological questions become especially relevant in this exercise, as divergent conceptions of forests have influenced different pathways of environmental rule, and definitions of the social and biological processes that are called deforestation and reforestation have shifted over time and among different actors.

Environmental rule is not just a power or governance problem—involving the state and subject peoples over contested resources—and not just an epistemological problem of how forests are defined and classified, but it is always an ontological problem as well (Carolan 2004). The physical properties and materiality of the world, or in our case, the lands and trees of Vietnam, have strongly influenced the policies and practices that are applied to them. Trees are not merely material objects acted on by people, but can be social actors in and of themselves. For example, specific biological properties of tree species (such as a demand for water or for being planted at certain distances from other trees in order to grow) may require distinct types of active labor for afforestation, which may in turn require certain types of policies and social relations. However, a focus on ontology does not mean that "nature" is a thing outside of human action: where trees can now be found in Vietnam has less to do with simple ecological factors of soil or climate and much more to do with social, economic, and political decisions of human actors. Trees and people together have remade each other in compelling and contingent ways through the interplay of power and politics, knowledge, and materiality.

COMPETING VIEWS OF FORESTS IN VIETNAM

What are the forests of Vietnam like, and how have they been understood? The answer to those questions varies across space and time; species common a century ago are now rare, and introduced species, like eucalypts and acacias that now dominate much of the landscape, were once nonexistent. Tree products fetching astronomical prices in the eighteenth century, like Vietnamese cassia (*Cinnamomum loureiroi*), can now be had for less than the price of a bowl of soup. Many households once dependent on local forests for food and fuelwood for subsistence now rely instead on planting global commodities like rubber or coffee for cash. The common theme for Vietnam's forests has been change.

Early adventurers to the Far East often remarked upon the scale and scope of Vietnam's forest resources, particularly the high number of economically valuable species in international trade, notably with China. A naval officer on the American brig Franklin, which pulled into Saigon harbor in 1819, wrote of the riches of the Cochinchinese countryside:

> The forests, besides the various kinds of odoriferous woods, such as the eagle, the rose, the sappan, and others, afford iron-wood, several species of the varnish-tree, the dammer or pitch tree, the gambooge, the bamboo, and the rattan, besides a great variety of woods useful in dyeing, in construction, and the mechanic arts. The country produces, also, cinnamon, honey, wax, peltry of various kinds, areka, betel, tobacco, cotton, raw silk, sugar, musk, cassia, cardomums, some pepper, indigo, sago, ivory, gold dust, rhinoceroses' horns, and rice of six different kinds. (White 1824, 248–9)

The allure of timber for navel shipbuilding attracted European attention, and, less than fifty years after White's initial description, the French established suzerainty over Vietnam's forests, justified not only by colonial demands for resources, but by concerns the forest estate was being wasted by native populations. Dr. Clovis Thorel, a French naval doctor with an interest in botany, wrote of his concerns:

> In terms of forestry all peoples of Indochina, including the Chinese, only know ways of destroying forests. Everywhere they burn forests, whether to grow forest rice, maize and cotton; to clear land so they can more easily move about and hunt animals; or, as we saw so many times, to simply distract themselves. This barbaric habit is general, and one can confirm that each year half of the total area of forestland in Indochina, even on many mountainsides, is ravaged by fire. This practice, so contrary to forest management in cold and temperate climates, is riddled with serious problems. . . . The practice of excessively stoking up fires, however, shows a genuine disregard for conservation which one cannot help but deplore the more one travels through the region. There are only a few rare places, distant from populations, where truly virgin forests grow in all their full power and grandeur, a state that forests so easily obtain in this climate. (Thorel 2001 [1873], 185–86)

Yet despite this plea that forest management needed to focus on native practices to ensure conservation of resources, a few decades later the French colonial forestry project was also to come in for critique. Well-known French geographer Pierre Gourou provided a frank assessment of the state of Vietnam's forests to colonial administrators, noting that more than fifty years of French rule, with its focus on private concessions for timber cutting, had done little to conserve resources and had in fact accelerated degradation in many ways, leaving forests "poor in valuable wood" and providing "not a centime of revenue" to impoverished residents near forests, given state and commercial control (Gourou 1940, 208).

Forests, degraded though they were by the mid-twentieth century, served another key function at the close of colonial rule: as important hideouts for resistance fighters during the more than thirty years of war that began in the 1940s. The revolutionaries' strategic use of forests for concealment eventually spurred the US military to spray herbicides in South Vietnam to remove vegetation cover along roads, canals, and other transportation arteries, as well as in dense jungles that were suspected of being supply routes for the Ho Chi Minh Trail. Over two million hectares of forests, ranging from coastal mangroves to high canopy dipterocarp forests, were subjected to herbicide attacks, including the defoliant Agent Orange, over the course of the war (Westing 1971). At the same time that US forces tried to systematically eliminate forests in South Vietnam, the need for a steady supply of wood for domestic use and wartime needs preoccupied North Vietnam. Forestry development complemented the process of collectivizing agriculture in the new socialist state, and building a new nation required building new forest institutions and practices as well. Large-scale State Forest Enterprises (SFEs) for logging were founded to provide the state with timber from wherever it could be usefully cut, and to encourage socialist labor practices in the rural mountains, and were extended to the south after the reunification of the two sides at the close of the US–Vietnam War in 1975.

The cessation of war brought peace, but years of isolation, until Vietnam again became the site of new attention to forests after international conservation organizations tentatively ventured back into the country in the late 1980s. The discovery of several new mammals previously unknown to science highlighted the fact that Vietnam's forest estate, while degraded, diminished, and fragmented, still held new forms of value. International donor aid flowed back into the country and was used to set up new nature reserves in previously overlogged forest enterprises. One of these new forest reserves was the Kẻ Gỗ Nature Reserve (KGNR) on the north central coast, established in late 1996 with a total area of 35,000 hectares, ostensibly to protect habitat for two endangered species of pheasant. Prior to this designation, Kẻ Gỗ had been the site of logging by four different SFEs, and was mostly composed of degraded secondary forests (in fact, Kẻ Gỗ in Vietnamese means "place with timber"). Approximately 40,000 people lived in the buffer zone of the new reserve, many of them migrants who had moved to the remote area as part of state development plans in the high socialism of the 1960s. These communities depended on forests to provide lumber for houses, fuelwood, charcoal, and non-timber forest products (NTFPs) that could be sold to supplement poor agricultural harvests in the area. Kẻ Gỗ, where I did a year of fieldwork from 2000 to 2001, was an ideal site for exploring the challenges of forest conservation and development in Vietnam.

Upon arriving in the fall of 2000, I met local village headmen to introduce myself and share the proposed schedule of my research with them. One of my very first conversations in the village nearest the nature reserve was with the elderly Uncle Thông, who had been sitting outside the headman's house, waiting for a meeting on party issues. With his close-cropped grey hair, olive-green army jacket, and pith helmet with the red star emblem of the Socialist Republic of Vietnam (SRV) on it, he looked the very model of a supportive citizen-subject. He inquired about what I was researching, confusing my presence with those of the foreign conservation groups that had visited before to help set up park activities. Despite assuring him that I was in his village to objectively understand the impact of forest policies on local peoples, he remained skeptical. "Hmmm," he harrumphed, and pointed at me, "You want to make a 'civilized environment' [*môi trường văn minh*], but we here are the ones who have to implement it." I asked him to explain further what he meant. He began to speak about being excluded from the forest reserve since it was demarcated a few years earlier in 1996, a process that had involved deployment of new government rangers and fences for boundary marking, despite the fact that local people were doubtful the endangered pheasants could even be found here, given the degraded nature of forests. Using the small bamboo table where we were drinking tea to represent the area of the Kẻ Gỗ forest, he began to move the little teacups around to represent villages that had lost access to the reserved area. Uncle Thông argued that relinquishing old lands and focusing on new environments took getting used to, as if a village were moved to Germany, at which point he abruptly sent a cracked cup off the table to a chair by his side. "This village," he said, pointing at the German teacup, "would have to learn a new language, new customs, a new environment to survive. Well, it's the same here, even if the distance is not." I nodded and asked whether he had raised these concerns about loss of forest access and social identity with authorities, since he was a Party member and highly involved in village issues. He shook his head no. "Don't get me wrong," he said emphatically, "I'm enthusiastic about protecting the forest (*phấn khởi về bảo vệ rừng*), but I want to emphasize that this forest is also for the livelihoods of the people, as well as for birds and animals."

Uncle Thông was an ambivalent subject of the new forms of environmental rule that had been applied to the forests of the Kẻ Gỗ area, and his uneasiness with the type of conduct that he should be involved in, such as supporting demarcation of state forests to protect a pheasant no one had ever seen, reflected many people's uncertainty regarding care of the environment. The concerns Uncle Thông expressed, that the conservation of forests, while presented as justified on ecological grounds, had required great sacrifices from local people and involved degrees of social engineering that citizens long accustomed to top-down state policy found uncomfortable, can be seen across multiple times and places in Vietnam. Contrasting and competing visions and histories of forests as valued, but endangered; as occupied by local peoples, but in need of protection from them; as rich in resources, but poor in economic benefits, have stretched across different generations, and have influenced how environmental rule has emerged.

THEORETICAL GROUNDINGS

Given my focus on understanding forest socio-ecologies, this study falls clearly into a field of inquiry known as political ecology. Political ecology aims to produce nuanced descriptions of resource use that go beyond simple biophysical explanations for environmental change, and instead looks at social influences and practices at multiple scales (Blaikie and Brookfield 1987; Blaikie 2008). For political ecologists, "nature," like capitalism or socialism, is not a fixed real entity; it is a concept of the imagination that must be conceived and described by actors from their being in and of the world (Ingold 2000). Thus it is the task of political ecologists to understand "the underlying processes through which particular global assemblages of nature and society are produced" (Braun 2006, 647). Authors working in this field have used a number of heuristic devices to refer to these complex socio-natures that combine nature and culture; for example, geographers have often called these assemblages hybrid landscapes (Robbins 2001a; Whatmore 2002; Zimmerer 2000). Other recent work has focused on the forms in which nature/culture combinations have been realized, such as in discussions about "technonatures," "post-humanist natures," or "multi-species interactions" (Braun 2004; Haraway 1991; Kirksey and Helmreich 2010). Perhaps the most prominent expression of this view is the recent declaration that we are now living in the Anthropocene, a geological age under total human influence (Zalasiewicz et al. 2010). Complementing these approaches, this book's case study of environmental rule at work shows the implementation of policy as complicated by ideas, identities, and physical properties of the environment that combine to result in formations that cannot be clearly identified as either exclusively social or natural.

ENVIRONMENTAL RULE AND GOVERNMENTALITY

My characterization of environmental rule as a unique but understudied phenomenon, and a key process in the creation of nature-society hybrids, owes much to Foucault's work on "governmentality," which first turned attention in the study of politics from the sources of sovereign state power to "relations of knowledge, authority and subjectivity" (Miller and Rose 2008, 5). Foucault believed that government was essentially a way to influence the "conduct of conduct," in his well-known formulation. This change from previous studies was remarkable; power had long been seen as that which was wielded by the state monopoly on violence, in Weber's phrasing, a kind of "negative, exterior" power (Mitchell 1991, 93). After Foucault, government could instead be seen as interior and productive, "a domain of strategies, techniques and procedures through which different forces seek to render programs operable, and by means of which a multitude of connections are established between the aspirations of authorities and the activities of individuals and groups" (Miller and Rose 2008, 61). It was these diffuse forms of power which Foucault famously labeled governmentality (Foucault et al., 1991).

By turning away from studies of states and regulations and towards knowledge, networks, and practices, a more complex, realistic, and contingent picture of governance emerges (Dean 1999). Yet applications of governmentality to the environmental field have remained somewhat more limited.[5] Environmental rule provides a clear example of how those seeking to change the "conduct of conduct" around the natural world have needed to engage new forms of authority, drawing less on power and laws, and more on knowledge, norms, and cultural approaches. A focus on environmental rule asks us to look not just at the means and measures by which environmental action is encouraged and enforced in specific policies and projects, but at the practice of politics in how problems like deforestation come to be recognized and defined in the first place, and the diffuse and networked actors involved in these processes.

Governmentality scholars have outlined how we might analyze the practices that make up everyday forms of political rule, although there has been some disagreement as to the methods that should be used.[6] Four specific aspects are of primary concern: problematization, knowledge-making, directing conduct, and making subjects (Dean 1999; Oels 2005). Each of these processes is also essential to understanding how environmental rule is established, in Vietnam and elsewhere, and study of environmental issues can evolve from these steps.

PROBLEMATIZATION: HOW ARE "ENVIRONMENTAL" PROBLEMS IDENTIFIED AND MADE VISIBLE?

As many governmentality scholars point out, the creation of a policy problem is a process that is not self-evident; we need to understand which questions rise to the point of being defined as in need of a solution, while other issues of concern may never be characterized as a problem at all (Miller and Rose 2008). Processes of problematization under environmental rule are diffuse, involve multiple actors, and often include the need to make things more "visible."[7] However, in the process of making an issue visible, in creating a "problem," often much contingency and complexity must be erased into order to conveniently bound the possibilities of solving it. Thus an important part of power that enables environmental rule, but one often overlooked in traditional policy analysis, is knowing what problems to frame and what to exclude (Li 2007). Knowledge production therefore often seeks to package the complex, messy, controversial, or contingent as anything but, and this process of weeding out complications and tidying up the problem to be solved has gone by many names in the wider literature on environment and development.[8]

How does problematization happen in environmental rule? Making a component of nature, such as forests, visible, to configure their existence and use as a "problem" in need of a solution, requires three things: nature *must be defined as an object of intervention*, often in ways that may be highly controversial and contested, but which are glossed over as universal, "scientific," or common-sense; these environments to be intervened in *must be visualized*, usually through the production and circulation of maps establishing authority; and *processes of change in these environments must be named*, as in directing attention to "deforestation," while other types of change are ignored. These practices taken together set the stage for later detailed interventions in conduct.

For this case study of forests in Vietnam, as a first step forests must be defined as objects of action and classification and enumeration of what constitutes a forest in a particular setting must be carried out. While it might at first seem easy to identify an object as a forest, to make it visible, it is in fact a complex procedure. There are standardized definitions of "forests"; for example, the Food and Agriculture Organization (FAO) of the UN defines forest as "land with tree crown cover (or equivalent stocking level) of more than 10 percent and area of more than 0.5 hectares" (FAO 2000). However, even standard definitions contain points of ambiguity, such as "whether various types of tree crops are to be defined as forest, as well as the point at which fallow in regeneration comes to be classified as forest" (Fairhead and Leach 1998, 8). Definitions and classifications of "forest" thus often come to be based primarily on political or economic factors (Peluso and Vandergeest 2001; Agrawal 2005). Trees themselves contribute to the problem and create ecological messiness—different plants grow various types of woody material, to different heights, in different places—which complicates attempts at classification. Typologies of forests, such as eco-types or biomes, are hard to standardize: some classification systems rely on leaf type (deciduous or coniferous), climate (temperate or tropical), physiognomy (old growth or secondary growth), or dominant plant type (Douglas fir or teak forests). There is no global standard classification system in place, and many countries use idiosyncratic systems based on local histories or usefulness of certain species.

In Vietnam, forest classification and reclassification happens regularly. From the colonial era, when the first lists of tree species were made, to the socialist era, which focused on timber-rich areas for the construction of a new nation, through to the modern era, when remotely sensed data show the carbon content of forests, classification has remained important. How forests are classified plays an extraordinarily important role in how they are later managed, and labels for certain forest types, such as those classified as "pristine" or "intact" versus "degraded" or "fragmented," then become rallying cries and justifications for specific forms of environmental rule (Forsyth and Walker 2008). The concept of "intact forest" carries with it associations of certain types of management (for example, often directed by the state), which would differ if we were to call the same plot of land a "tree farm," a "plantation," or a "field" (which might be managed by individuals). Thus classification and definitions of forest are crucial moments, from defining a forest worthy of protected areas status to definitions of what constitutes a "tree," and these patterns of classification have real material consequences. For example, designations of certain lands in Vietnam as "bare hills," even though these lands were being actively managed for other uses, turned them into targets of reforestation projects in the 1990s. These classifications ended up dispossessing poor families and women of sites for pasturage or fuelwood in favor of richer families who planted exotic eucalyptus for pulp factories (chapter 4).

Next, forests, once they have been defined, must be made visible, usually through mapping. Maps are key to environmental rule: management regimes, land tenure politics, and funding practices all center on the idea that a forest must exist in certain designated areas if it is found on a map. The production of seemingly objective maps that represent biophysical reality is in fact a largely political process, as many geographers have long asserted (Kitchin and Dodge 2007). Within Vietnam, French authorities often made maps even when there had been no on-the-ground assessment of actual land types, labeling large green swaths of the countryside of Indochina simply as unexplored forests (*forêt inexplorée*). Later Vietnamese authorities strongly promoted one specific visualization of forests: a map made by a French forester that claimed that Vietnam had 43 percent forest cover in 1943 (map 4.01). Despite the fact that this map made clear there were very few areas of rich or primary forest remaining after years of colonial exploitation, the picture was later reproduced by authorities in the 1990s and became an important tool to guide a massive reforestation program. Even advances in mapping technology, moving from the hand-drawn cartography of the colonial era to advanced satellite imagery today, has not obviated the need to pay attention to how maps present certain views and not others, and the ways in which these images are contested (Robbins 2001b).

Finally, processes of environmental change must be defined, or framed, and some identified as problems. That is, deforestation must be named before it is noticed. When the processes that make up land use change happen, such as cutting lumber or burning shrubs for pasture, they appear to be less of a problem if they are not called "deforestation." Definitions of what counts as deforestation have shifted over time, and from actor to actor, providing rich opportunities to see knowledge creation and policy mobilization in practice.[9] Often, these discourses of deforestation could be more accurately described as a type of "environmental orthodoxy," as Timothy Forsyth has termed them; these orthodoxies are "institutionalized, but highly criticized conceptualizations of environmental degradation" that continue to circulate and influence real-world policy decisions, even when they have been shown to be inadequate, simplistic, or downright wrong (Forsyth 2002, 37).[10] All too often, these orthodoxies are constructed at a distance, and are aimed not at understanding the underlying dynamics of change, but how these changes might be halted or harnessed for already determined goals of improvement (Li 2007). These orthodoxies demonstrate how important the framing of problems is, and how crucial politics and power is in allowing certain framings to dominate and circulate above other alternative understandings; Maarten Hajer has termed this process the creation of "discourse coalitions" (1993). The new explanation we can add, via the lens of environmental rule, is that these narratives and discourses often circulate and become supported by coalitions because they are helpful in masking the true reasons for so-called "environmental" interventions.

Within Vietnam, contemporary concerns over deforestation often focused on the idea that forest change in the late twentieth century was rapid, extensive, and linked solely to human population movement and growth. Yet even in the nineteenth century, French authorities bemoaned the lack of primary forests on their arrival, and re-imagined vast areas of Vietnam as dense forest that must have been lost to deforesting natives in the recent past. In reality, deforestation has been ambivalently addressed by the state and by citizens, with some deforestation seen as necessary by policy makers (for example, state logging) and other types of deforestation in need of reform (such as ethnic minorities practicing swidden agriculture). How certain forest use practices became problems, and coalitions and networks built to tackle these problems, while others were ignored, is taken up in each of the chapters of the book, as these problematizations have changed over time.

KNOWLEDGE-MAKING: HOW IS ENVIRONMENTAL 'TRUTH' CREATED AND USED?

Governmentality and environmental rule are both fundamentally undergirded by knowledge production. Scholars have placed emphasis on the fact that knowledge is not, as positivists have long claimed, a way to render the real world understandable in objective ways; rather, knowledge production makes an object real, and thus "susceptible to evaluation, calculation and intervention" (Miller and Rose 2008, 66). Ideas about the world and its problems are produced and co-produced by multiple actors, and often require particular technologies in order to be conceptualized; sociologist Michel Callon has termed these tools "*dispositifs de calcul*," or calculative mechanisms, that render or translate newly visible problems into discrete parts amenable to solutions (Callon, 1998). We can think of the many tools by which much early state authority operated, for example, which were in fact knowledge tools and calculative mechanisms: maps to fix borders and boundaries, geological explorations of the best soils for cash crops, or the use of statistics to represent forests (Agrawal 2005; Demeritt 2001). It also becomes important to have information about forest users via similar tools. Classification of peoples, particularly indigenous or marginalized ones, and landscapes often operate in tandem; it is not a coincidence that at the same time as forests have been studied and mapped, so too have nearby populations become the object of ethnographic and spatial gazes (Sivaramakrishnan 1995, 2000; Bose et al. 2012). All are mechanisms of calculation that enable knowledge to be turned into interventions: moving populations, taxing them, or asking them to replant forests and protect species. Environmental rule depends on these calculation techniques because policy often becomes less controversial and less political if it can be cast in the dry languages of data and science. For example, a map that shows plans to confiscate land from households would quickly arise ire and protest, but a map showing the location of endangered species that need to be protected (and which ignores the social impacts of such intervention by leaving out settlements or people), might be far less controversial.

Further, production of knowledge often requires, and indeed forms the parameters of, expertise, and governmentality scholars have documented the concurrent rise of experts within the modern problematics of government (Mitchell 2002) as well as documenting the authority that this expertise creates (Li 2007; Mathews 2011). Experts are regularly called upon to help enact environmental rule, and often, expertise and authority become intertwined as ruling actors emerge. Others have written of the "performativity of data" (Waage and Benediktsson 2010): to have statistics, reports, inventory data and the like is to be able to perform this expertise. "Science" is regularly invoked in the construction of expertise for rule, usually with an eye to asserting the apolitical and objective nature of scientific inquiry. Yet the practice of science itself is a form of politics, and it needs to be incorporated in any analysis of policy outcomes (Jasanoff 2004; Forsyth 2011). It is also important to recognize in any analysis that knowledge production under environmental rule is never simply top-down from experts, but is often multifaceted, as it is influenced by co-production: local knowledge can have as important an effect on the way rule is enforced or resisted as formal, "scientific" knowledges (Birkenholtz 2009; Peluso 2003).

We can see how these processes play out in the production of knowledge for environmental rule for forests. First and foremost, *forests must be given value through expert knowledge and mechanisms of calculation*. Specialized training in forestry grew throughout the nineteenth century, centered on Germany and France (Scott 1998), and spread throughout the world via colonial management techniques and eventually the emergence of a global institution, the Food and Agriculture Organization, where "expert" foresters provided calculative techniques for timber production, smallholder management, and other approaches (Vandergeest and Peluso 2006b). Mechanisms of calculation for forest knowledge have included forest mensuration inventories to determine species, age, volume and form of timber-producing trees; test plots to measure regrowth after cutting; working plans to rotate cutting throughout a forest to even out age classes; studies of the hydrological impacts of forest cover; and genetic manipulation and breeding of tree species (Galudra and Sirait 2009; Mathews 2005). Through these techniques, "expertise" about forestry becomes privileged; that is, specialized training in forestry as a science and discipline becomes valued over local knowledge of actual forests in specific places.

Knowledge-formation about forests in Vietnam has focused on a few key themes, but what they share in common is a focus on knowledge that is generated by elites, and then internationalized and articulated by formal institutions. French colonial forestry research, explored in chapter 1, dwelled on the links between forests, hydrology, and slope, influenced substantially by the experience in the metropole with Alpine floods. Knowledge formation about forests in Vietnam has always been highly internationalized, even when economically cut off from much of the Western world at the close of the US-Vietnam War. Economic bedfellows in the socialist bloc, namely Eastern Europeans and Russians, provided training and research on forest composition that was developed in the cold temperate regions of Europe, and which was not always relevant to the tropical diversity of Vietnam (chapter 2). The arrival of multilateral donors and global conservation NGOs in the past twenty-five years has resulted in funding shifts to focus on species biodiversity and landscape conservation (chapter 3). The most recent approach to knowledge generation has reverted back to a focus on forests for climate regulation, as new attention has focused on the idea of forests as providers of "environmental services" (chapter 5).

Expertise has shifted over time as well, as environmental rule is characterized by diffuse knowledge and discourse coalitions that are often temporarily strong, but unstable. From colonial foresters who were required to be trained at the metropolitan forestry school in France, to contemporary forestry officials who need to be fluent in a bewildering array of acronyms for participation in international carbon projects, expertise has varied. So too have varying concepts of forest ("*rừng*") meant different things to different actors, ranging from committed Marxist officials working in upland environments, international NGO staff, and forest farmers. Local knowledge is therefore explored throughout the chapters of the book as well, particularly as it interacts with and influences other forms of "expert" knowledge.

INTERVENTION: HOW IS ENVIRONMENTAL CONDUCT SHAPED?

Technologies of knowledge-production often seep into technologies of rule and conduct-shaping and the two cannot be easily separated into distinct categories, as "the activity of problematizing is intrinsically linked to devising ways to seek to remedy it. So, if a particular diagnosis or tool appears to fit a particular problem, this is because they have been made so that they fit each other" (Miller and Rose 2008, 15). This firm link between the shaping of knowledge and the shaping of conduct is key (Li 2007). Thus a study of the spatial distribution of endemic species may lead directly to a policy of reserving certain lands for nature preserves, or a cadastral map that shows where populations are leads directly to a policy to resettle villages. A focus on knowledge about forest hydrology may lead to restrictions on tree use in upland areas, rather than limits on water use in downstream areas (Forsyth and Walker 2008).

The danger inherent in environmental rule is that by focusing on the "environmental" aspect of interventions, when in fact social interventions are actually envisioned or planned, this sets into motion a process that may lead to potentially negative outcomes: a misidentification of actors affected by interventions (for example, attention to ecology or biology but not people); a lack of participation by those most impacted; and other aftereffects. For example, a policy on environmental services that is limited in vision and intervention to monitoring hydrological flows, but not the livelihood impacts of water use by actual citizens, risks missing out on the bigger picture of how people and water mutually interact. This is just one of the potential pitfalls of contemporary environmental rule.

Interventions are not simply directed by states and other formal authorities. Increasingly, governance in the modern era is limited by modest finances, decentralized institutions, and perceived freedoms, or what Foucault called "frugal government" (Foucault 2008, 322). Newer neoliberal approaches to governance (that is, ones that rely on markets rather than states) have meant that interventions are increasingly shaped at a distance or by weakened governments, but even these neoliberal approaches similarly depend on calculative practices leading to interventions to encourage particular conduct. For example, attention to trees as stocks of carbon and regulators of water has been used in the past five years in Vietnam to create a new policy on payments for environmental services, which transfers cash to households who volunteer to protect forests, indicating that environmental rule does not always take place through disciplinary power, but through diffuse forms such as markets and voluntary initiatives as well (chapter 5).

Yet to shape conduct in any form requires change, and change is often resisted by both nature and society. It is thus important to stress that "technologies of rule will not always achieve their stated effects" (Corbridge et al. 2005, 6), and resistance in various forms to interventions of rule becomes an important component of how rule develops.[11] Interest in resistance has been formative in the field of political ecology to understand the ways in which technologies of rule fail to persuade the governed to conduct their conduct differently (Guha 1990; Peluso 1992; Scott 1985). In Vietnam as well, resistance to environmental rule has taken different forms, from subtle foot-dragging and professed local "ignorance" of new forest restrictions under French colonialism (chapter 1), to active smuggling and violent protests against forest guards in more recent years (chapter 3). Sometimes this forest resistance achieved goals of remaking environmental rule, and in unexpected ways. For example, counter-conduct against forest restrictions in the 1930s played a large role in the emergence of the Indochinese Communist Party, which would later transform the entire nation of Vietnam.

SUBJECTIFICATION: HOW ARE SUBJECTS OF ENVIRONMENTAL RULE FORMED?

A final key question to understand environmental rule is identifying "what forms of person, self and identity are presupposed by different practices of government and what sorts of transformation do these practices seek?" (Dean 1999, 32). Individual subjectivity becomes an essential mechanism by which rule is enacted, through the myriad ways people consider, evaluate, understand and act on the things they are "supposed" to do. Examples within environmental policy include decentralized forest management to encourage responsibility among a previously forest-encroaching peasantry (Agrawal 2005); involving individuals in biodiversity conservation through affective forms of action like bird surveys and conservation volunteering (Lorimer 2008); and encouraging citizens to monitor their own carbon emissions (Paterson and Stripple 2010). Arun Agrawal's exploration of subject formation in India's forests has noted that these "new environmental subject positions emerge as a result of involvement in struggles over resources and in relation to new institutions and changing calculations of self-interest and notions of the self" (Agrawal 2005, 3), and he has termed this merging of governmentality with subjects coming to see the environment as an object to be cared for as "environmentality."

In Vietnam, successive waves of environmental rule have encouraged logics of care and responsibility around forests, but these have not always succeeded; unlike Agrawal's case in India, environmental rule does not necessarily lead to environmentality (that is, to caring about the environment). As with resistance to techniques to shape conduct, so too has there been resistance to new subject formation. Other scholars have pointed out conservation activities do not always result in reshaping of consciousness in desired ways, as projects are often constrained by questions of exchange and power (West 2006). One additional reason for resistance may be that subjects of environmental rule can often tell when a so-called "environmental" intervention has other, often more inconsistent or exploitative goals, particularly those that might benefit rulers more than others, and as such, they see through calls to behave in environmentally-friendly ways.

Making environmental subjects in Vietnam has required changing discourses around both citizenship and nature. Subjectification has taken many forms, and contrasting the acceptance of some new environmentalities with the rejections of others gives us insight into this process. Perhaps the clearest link between the environment and subject-making occurred during the early years of state socialism, when state authorities set up collective forest farms and logging operations. In addition to providing the war-torn nation with timber, these cooperatives also served to create new socialist worker-subjects and eliminate familial and ethnic ties that might subvert this new identity (chapter 2).

ENVIRONMENTAL RULE AND ACTOR-NETWORK THEORY

While governmentality theory helps us form the basis for understanding key concepts of power, knowledge, conduct and subjectivity in environmental rule, it does fall short in explaining how both natural and social forces mutually help constitute one another in processes of co-production. This is no doubt due to the fact that Foucault had surprisingly little to say about nature and non-humans, despite his common phrasing of governmentality as being about the "imbrication of men and things" (Foucault et al. 1991, 93–96).[12]

One field of theory that does take objects seriously in their roles and capabilities is actor-network theory (ANT), and ANT gives us a more complete picture of how environmental rule happens than governmentality alone. Arising out of the larger field of science and technology studies (STS), ANT is less a full-fledged theory and more a series of observations about the ways in which knowledge is created and circulates and how subjects and objects of this knowledge cannot be separated out into opposing realms. ANT can be useful, when combined with governmentality approaches, in seeing how environmental rule arises and circulates, as it helps us understand how rule can emerge even in the absence of a centralized directing state.

Based on his early studies of laboratory scientists such as Louis Pasteur, Bruno Latour has argued that all science practice is local, and it is in fact relationality (not scientific "objectivity") that creates both knowledge and its circulation (Latour 1987). This places emphasis on the networked aspects of knowledge production, and the need to understand the who and what comprising the network. Actors themselves are key, as they "become powerful through their abilities to enroll others in a network and to extend their network over greater distances. Building networks depends on actors' capacities to direct the movement of intermediaries such as texts, technologies, materials and money" (Burgess et al. 2000, 123–4). ANT scholars have emphasized that the power of knowledge itself is often less important than the power of networks to promote new ideas through a process of "*interessement*," by which some of the networked actors are "made interested" and then potentially become "spokespeople" for the new idea/thing (Akrich et al. 2002). A mobilized, interested, enrolled network emerges through this process of "translation," as ANT scholars term it, to generate knowledge and action about some scientific problem (Law 1992), whether it is how to regenerate scallops in a French fishing community (Callon 1986) or how to replant forests in denuded areas of Vietnam.

What forms do these myriad networks take, and how might they relate to environmental rule? Some authors have looked at commodity-chains as networks, in which timber, minerals, food or other objects are harvested, acquire value, are exchanged, and which make new forms of economies at different scales (Jarosz 2000). Others have shown how citizen-science activities, like participatory bird counts, enroll new subjects in caring about endangered species, and that networks of scientists, citizens, birds, biodiversity checklists, and habitats come together to shape biodiversity policy (Lorimer 2008). As these examples indicate, both objects and humans within networks are considered potential actants: as Latour has noted, "In addition to 'determining' and serving as a 'backdrop for human action,' things might authorize, allow, afford, encourage, permit, suggest, influence, block, render possible, forbid, and so on" (Latour 2005, 72).[13] Nature is not an abstract concept in this visioning, but a real material thing that often "pushes back" and can derail, reroute, or expand the original conceptions of actors pushing certain forms of environmental rule.

As part of environmental rule, objects arise and circulate through relational human interactions, but these interactions are shaped by the physical, biological, and social characteristics of the objects themselves. Networks of human and nonhuman come together in the production of socio-natural landscapes, and the forms of environmental rule that evolve to manage them both. Forests thus become an effect and outcome of these networked processes. Much like a pinball may never take the same path twice after being launched by a plunger, tracing the network itself allows us to see the ways in which environmental outcomes are affected by pathways, actors, events, objects and effects operating in modes of constant change. It is this networked aspect of environmental rule that gives it such dynamism.

Networks shift and change under environmental rule, and need to be studied accordingly. During the colonial era, the small numbers of people allowed to be official foresters (namely, French men who had studied at the National Forest School in the countryside of Alsace-Lorraine) sustained a small network between Indochina and the metropole of like-minded people concerned about hydrological issues and deforestation, and who tried to enroll other actors, such as shipping merchants, local mandarins, and hydraulic engineers, into actions supporting forest reservations (chapter 1). During the socialist era, networks of forest engineers and wood technicians circulated among friendly Communist bloc countries, advising on how to maximize biological production in line with Marxist principles (chapter 2). Later, more expansive networks to support afforestation efforts have included international donors and their reports, documents, and projects on forest loss in Vietnam; government departments, with laws, statistics, and state officials promoting certain types of fast growing seedlings; local villages and households, in their actions of replanting or cutting forests; and the physical landscapes themselves, with trees moving around in space and time, growing unbidden in some areas and requiring considerable labor in others (chapter 4). Often these networks have been loosely connected and ephemeral, yet influential.

Forests themselves as objects engage in networks of environmental rule in different ways. Anthropologists have long been interested in the deeply social and symbolic roles that trees and forests have played in multiple cultures, from tree worship ceremonies and sacred groves to myths and stories of place (Rival 1998; Jones and Cloke 2002), and increasingly attention is being paid to trees as actors in social networks as well. Trees in forests grow at different rates; they produce different types of more or less valuable wood; they burn down or refuse to grow; and they spread out into new territories, often without human assistance (Jones and Cloke 2008). Additionally, forests and other ecologically defined natures often serve as a linking concept, or "boundary object," that joins disparate actors in a network.[14] For example, the abstract, weakly defined idea of "reforestation" in Vietnam was able to enroll stakeholders as diverse as state officials interested in expanding wood processing industries and international NGOs concerned about biodiversity to work together to support afforestation efforts (chapter 4).

Forests themselves are not the only objects enrolled in networks of actors that shape environmental rule. So too are the material forms rule takes, such as land tenure certificates, resettlement sites, or markets for environmental services, as these objects create certain types of action while constraining others. For example, the French colonial forest service often posted signs and regulations to indicate where approved forest activities could be carried out (chapter 1). But these written posters often served as a material target that locals could protest against in ways that would have been more difficult and diffuse without such objects. Treating the technologies of rule themselves, whether they are labor camps for state logging or exotic seedlings for reforestation, as mobile objects with certain types of agency allows us to better see the networks, flows, and relations that make up environmental rule on-the-ground. Understanding the different paths by which these objects arose and were represented also helps us see why controversies over the outcomes of environmental rule can also emerge in unexpected ways.

THEORIZING FROM ENVIRONMENTAL RULE

Combining theoretical guidance from both governmentality and ANT points out the contingent nature of environmental rule: how it occurs, how it is viewed and represented, and how change is responded to. Linking theoretical approaches also helps us understand the "why" of environmental rule: why are environmental justifications used for certain policies that are in reality social or economic, and when? In some cases an environmental goal may be chosen simply because it seems less controversial than a social one. As Stephanie Rutherford (2007) notes, "The saving of nature is often taken for granted as an innocent endeavor, never implicated in relations of power and a noble exercise for the good of all life" (295). In other cases, environmental rule may grow in not-well directed ways from existing networks (for example, rule may evolve out of mundane mission statements, whereby certain organizations may feel they must cast all their projects as "environmental," no matter to whom or where they are directed). In other cases, more nefarious explanations may be afoot, in that casting policies as "green" might help to cover up true goals that enrich elites or dispossess the poor. There are examples of different underlying justifications for rule throughout this book, ranging from the concerns over flood control that the French forest service often used as justifications for expanding economic reserves for timber, to the professed concerns over biodiversity that were used by officials in the 1990s to obtain donor dollars for degraded and denuded forests that held little in terms of conservation value but which employed thousands of people in state jobs (chapter 3).

We also need to understand the "how" of environmental rule, particularly in terms of outcomes. Many interlocutors in Vietnam often expressed disappointment to me about the many failures of forest policy over the years: failures to halt deforestation and illegal logging; failures to prevent raw log exports and smuggling across borders; failures to meet afforestation targets; and failures to protect biodiversity. Coupled with these outright failures were unintended consequences of forest policy as well: poorer households receiving less forest land allocations than richer ones, or forced resettlement of households from protected areas leading to impoverishment. One explanation for these failures is that they were inevitable outcomes of systems of environmental rule that had identified the wrong problems, generated the wrong knowledge, intervened in the wrong places, and targeted the wrong people. Yet the worst and most negative outcomes were not always deliberate or underhanded. Rather, missteps in problem identification or knowledge production may have led to inadvertent focus on actors that were not important, or ignoring those who were significant. Outcomes of environmental rule are thus hard to predict in advance, which makes empirical historical-ethnographic studies such as this one all the more important.

The chapters that follow attempt to answer the "why" and "how" questions surrounding environmental rule by looking at how forests have been characterized in Vietnam and the directions that were given to shape their use and management over time. The book is arranged in a chronological fashion, starting from French colonialism through the high socialism of the 1950s through the 1970s, transitioning to decollectivization and decentralization as Vietnam moved toward a more neoliberal economic model, and

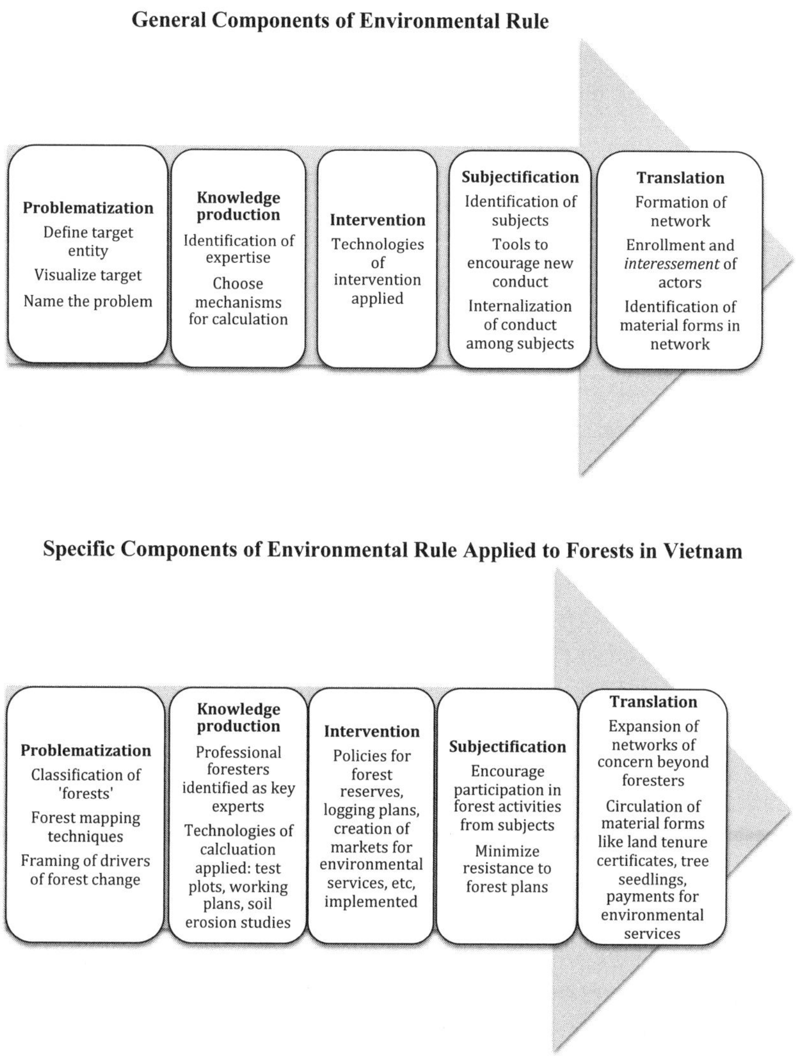

General Components of Environmental Rule

Problematization
Define target entity
Visualize target
Name the problem

Knowledge production
Identification of expertise
Choose mechanisms for calculation

Intervention
Technologies of intervention applied

Subjectification
Identification of subjects
Tools to encourage new conduct
Internalization of conduct among subjects

Translation
Formation of network
Enrollment and *interessement* of actors
Identification of material forms in network

Specific Components of Environmental Rule Applied to Forests in Vietnam

Problematization
Classification of 'forests'
Forest mapping techniques
Framing of drivers of forest change

Knowledge production
Professional foresters identified as key experts
Technologies of calcluation applied: test plots, working plans, soil erosion studies

Intervention
Policies for forest reserves, logging plans, creation of markets for environmental services, etc, implemented

Subjectification
Encourage participation in forest activities from subjects
Minimize resistance to forest plans

Translation
Expansion of networks of concern beyond foresters
Circulation of material forms like land tenure certificates, tree seedlings, payments for environmental services

FIGURE I.02 General and specific aspects of environmental rule. Figure by author.

eventually to a internationalized forest policy in which the nation's trees were imagined as global carbon storehouses. Although this may imply a consecutive trajectory, in reality, these pathways are rarely straight; rather, multiple pressures and questions about who should have rights to forests and for what uses have played out in different ways at different points in time. Shifts in environmental rule from more disciplining forms of domination, such as threats of state violence that characterized the French colonial period and early postcolonial rule, have given way to more diffuse forms of power in the neoliberal era, involving new intermediaries between the state and citizens, and through practices of economics and markets rather than control and punishment. Looking at environmental rule as a practice, rather than as a political ideology alone, also helps us see where there are continuities between what appear to be radically different periods of time.

While Vietnam is the basis of this case study, the methods used can be applied elsewhere. Each chapter examines the interrelated components of environmental rule: how were forests, forest change, and people *defined as problems* at different times, such as in the idea of "deforestation"; how was knowledge about forests and forest-using communities *generated, circulated, and acted upon*, and by whom; what technologies and tools were used to try *to shape interventions and conduct* in forests; how were *different subjectivities formed* through these myriad practices and approaches to policy; and *what modes of translation occurred* as ideas, practices and material goods moved through networks associated with different forms of rule? (See figure I.02).

The book concludes with the argument that understanding environmental rule helps us reject simple explanations for environmental change, as no single structural narrative works to explain the overarching processes of the transformation of material nature and human subjectivity over the long twentieth century in Vietnam: not Malthusian population growth, not capitalism, not Marxism, not neoliberalism. Instead, multiple actors intersecting with each other and with material objects like trees shaped the politics that emerged regarding how people and landscapes might best be managed at different points in time. Multiple drivers of change have often operated at the same time, helping us explain anomalous outcomes, such as how deforestation and afforestation could both be on the rise simultaneously. Detailed empirical studies like this book, which engage in ethnography and history, as well as attention to nonhuman natures, thus improve our understanding of the ways in which our world is remade, again and again, through multiple forms of environmental rule.

1 Forests for Profit or Posterity?

The Emergence of Environmental Rule under French Colonialism

COLONIALISM IN INDOCHINA OFFICIALLY BEGAN WITH EMPEROR Tự Đức's acceptance of France's claims on Cochinchina in 1862, and the later establishment of French protectorate status over Cambodia in 1867, Annam and Tonkin in 1884, and Laos in 1893. Among other changes, the colonial enterprise brought new forestry laws and management practices to Indochina, which played important roles in the transformation of the human and natural environment. Before it was possible to cut timber for railways and telegraph poles for the expansion of the empire, it was necessary to define those spaces in which this extraction would be allowed to occur, and where it would not.

Environmental rule thus first emerged from the colonial state and relied on the problematization of forests as depleted due to native practices and in need of protection and management organized by the new authorities. As French administrators set up the first forestry institutions, they spoke often of environmental concerns, particularly regarding the links between forests and water supplies, but in reality, other justifications dominated state forest policy. These included the need to fund the forest service through leases and taxes on private concessions, given the stingy funding of the colonial enterprise by France, and the need to control and settle native populations, given worries that opposition to forest regulations might coalesce into opposition to French rule generally.

This process was quite typical of the Southeast Asian colonial experience, as scholars have identified this era as one of "political forests," or the designation of forests based on political and economic criteria to be managed as a responsibility of the state (Peluso and Vandergeest 2001). This is not to say that the state was always an active forest manager itself, but rather that the state set the boundaries of the definitions of forests, determined the actors who could and could not use forests, and arbitrated disputes that might arise between actors. As elsewhere in Asia, the colonial period first codified

Indochina's forests into an entity to be controlled by the state via techniques of intervention, such as spatially delimited territories and economic regulation through taxes, permits, and concessions (Buchy 1993; Thomas 1999). New forest controls attempted to limit native harvesting by cracking down on practices deemed especially destructive, such as the use of fire and shifting cultivation.

As in all projects of environmental rule, knowledge was key to the imperial forest mission. French foresters arriving in the new colony problematized unorganized and unscientific forest use as the main cause of degradation. To control these local practices, knowledge had to be produced about both landscapes and people. New botanical gardens and scientific research institutes emerged and catalogues of species and maps of forest cover were composed. Studies also had to be initiated on how to apply locally the professional forestry practiced in France, as experiences with temperate, Alpine, and Mediterranean species were of little use in the tropical world. Analyses were done of the local climate services provided by forests and their linkages to downstream water flow; these took on an added importance after 1915 when a flood on the Red River nearly wiped out Hanoi, the capital city of Tonkin and seat of French colonial rule. The people in forested areas, both Vietnamese and ethnic minorities, were similarly studied, particularly their use of fire to burn forests for agricultural planting in swidden fields and to create pastures for livestock, as this was considered a major driver of deforestation, not to mention a waste of timber that could be providing revenue for the colonies.

The proposed solution to the myriad woes of poor quality forests, low timber yields, and undisciplined forest use by local people was new state interventions: the establishment of forest reserves where exploitation could be rationalized. New forest regulations would be codified and physically posted in villages to restrict local forest use to certain stands and to keep populations out of more valued reserves for which the French forest service would have exclusive authority. Yet these "Reserved Forests," though often justified on environmental grounds, were not to be conservation areas where ecological protection could be assured. Rather, they would be subject to methodical treatment through rotational cutting plans, and rights to timber sold for public auction by private contractors, as was done in state forests in France, with tax revenues going to colonial authorities.

Yet these extensive knowledge practices and interventions in local conduct failed to induce locals to change their forest use practices or produce compliant colonial subjects. Authorities could not establish extensive state-managed forests due to low personnel levels and the late and contested nature of France's presence in the region. And unlike the case with other French colonial holdings, particularly in Africa, ecological conditions in Indochina also prevented ambitious plans from being realized: the difficulty of logging and the highly diverse ecology of Indochinese forests kept forest administrators continually worried about financial losses. Even this somewhat lax forest management system did not create new forest subjects willing to engage in conservation, but rather alienated many locals who were subjected to new controls, such as limits on the collection of fuelwood or duties and taxes imposed upon timber. These access restrictions, and the local protests that accompanied them, played an important role in sparking rural unrest that was directly tied to the emergence of the Indochinese Communist Party in the 1930s, which was to wrest back control of the country under the leadership of Ho Chi Minh. In this way, the era of French colonial forestry provides an illuminating lens into the birth of not only environmental rule, but also modern Vietnam itself.

FORESTRY FROM THE METROPOLE TO THE COLONY

To a large degree, colonial forestry in Indochina was shaped by experiences imported directly from the French countryside. There, fears of rapid deforestation due to expanding agriculture, construction, and commerce from the thirteenth to seventeenth centuries resulted in an Ordinance on Waterways and Forests promulgated by the king's finance minister Jean-Baptiste Colbert in 1669. The new ordinance contained no fewer than five hundred articles, pertaining to both royal and private woodlands, and "gave the king significant new rights over French forests and notably expanded royal prerogative" (Pincetl 1993, 81). Two primary components of the ordinance stood out: first, the requirement that each forest of a minimum size should be divided into compartments and felled in rotation with standards (seed trees) left behind. Second, the state enforcement of laws over forests was strengthened through a new centralized bureaucracy and criminalization of forest infractions (Bamford 1955). Knowledge production and standardization were fundamental to this emergent state forestry apparatus; for example, the new Ordinance defined common units of measurement for woodlands and even bundles of fuelwood, and required private wood-owners to make detailed reports of wood felling (Graham 1999).

Such efforts did not succeed in slowing deforestation, however, as "the effectiveness of forest law seems to have tended to decrease as the physical distance from Versailles increased" (Bamford 1956, 85). The post-revolution founding of the *École Nationale des Eaux et Forêts* (National School of Water Resources and Forestry) at Nancy in 1824 marked a significant turning point. So did a new forest law in 1827 that reiterated precepts from the 1669 ordinance, but which further consolidated formerly communal forests under state regulations. An additional impetus was massive floods in 1840 and 1855, which increased attention to the links between deforestation and water flow from upland areas.[1] Pointedly, foresters blamed the local peasantry living in Alpine areas as the cause of mountainous degradation and set about focusing forest restrictions in these areas, which not coincidentally also lay on the edges of the consolidating French nation (Whited 2000b).

TRANSPLANTING FORESTRY PRACTICES

These experiences from the French countryside must have been much in the minds of officials who arrived in the Indochinese colonies tasked with developing a nascent forest service. Unfortunately, we do not have a good understanding of the extent of forests or local management practices at the time the French first arrived, as there is no comprehensive assessment of environmental resources for the precolonial era.[2] Goods from forested areas had long been important in maritime trade, used as tribute linking the Vietnamese court to other imperial dynasties in the region (Werner et al. 2012).[3] The royal demand for certain forest products drove some early conservation regulations on species that the Emperor prioritized, such as hardwoods for coffins (Fangeaux 1931). Emperor Minh Mạng (r. 1820–41) also made the cinnamon trade from wild trees a royal monopoly, as it was highly profitable (Hickey 1982). There is no indication, however, that the early courts created any protected forest areas, nor attempted to restrict forest use beyond a few favored species. For example, there are no records of taxes or restrictions on forest land in the detailed Nguyễn dynasty chronicles such as the *Institutions of the Successive Dynasties* (Lịch triều hiến chương loại chí)[4] or the *Unification Records of Đại Nam* (Đại Nam nhất thống chí).[5] The royal court also did not attempt to restrict the use of swidden fields for agriculture among the non-Vietnamese populations in the highlands on the edges of the imperial domain (Tạ Văn Tài 1985).

Thus, when France claimed Indochina, it began to impose a forest regulatory system different than previous dynasties. Forest management proved an important concern from the very beginning, as the need for mundane technologies of colonial rule, such as telegraph wires and railway sleepers, necessitated much forest felling. In 1862, just months after the first provinces of Cochinchina fell to the French military, the governor, Admiral Bonard, issued a decree that certain trees (known as "classified species") were to be used exclusively for public works of road and shipbuilding (Guilliard 2010). At the same time as the military began increasing their demands for timber in Cochinchina, they noted their concern with local use of forests, complaining that:

> The responsibility of mayors, with respect to the conservation of wood of their commune, is absolutely nil, and since we know their extent and capacity, no serious control can be exercised over the cutting of trees, because of the scarcity of resources available to inspectors. . . . [A]ll previously existing associations for the preservation of such species have disappeared today, and the administration cannot rely on either community authorities or private interests to prevent unnecessary damage, fire without cause, or fraudulent operations. (Pierre et al. 1866, in Thomas 2000, 54)

In response to these and other concerns, administrators created a new Forest Service for Cochinchina in 1866, and in 1875 passed the first law on the protection of forests. Certain species were placed on regulated lists to prevent excessive harvesting (sixty-three species total had minimum felling diameters), and three forest reserves were set up in Thủ Dầu Một province where logging would be more strictly enforced (Thomas 2000, 58) (see map 1.01). The new law on forest protection also laid out rules for reforestation, and heavy fines for those who illegally logged or allowed livestock grazing in forests.[6] However, the actual number of people involved in the Cochinchinese Forest Service remained low, estimated at under a hundred by the turn of the century.

Paul Doumer's accession as the governor-general of Indochina in 1897 led to an overall centralization of many aspects of colonial life, including in forest management. A Forest Service (*Service Forestier*) was established for each division of the Indochinese Union, along with a local Forest Law, with a central overseeing office in the Directorate General of Agriculture and Forestry in Hanoi.[7] As was the case in nearly all French colonies, the new forest laws were based in part on the 1669 and 1827 Forest Codes from France, considered "the bible" of forest management (Madec 1997).[8] The laws in both the metropole and the colonies focused primarily on the state as land manager, the use of working plans for rotational felling for sustainable timber production, and strong criminal provisions for violations of the law.

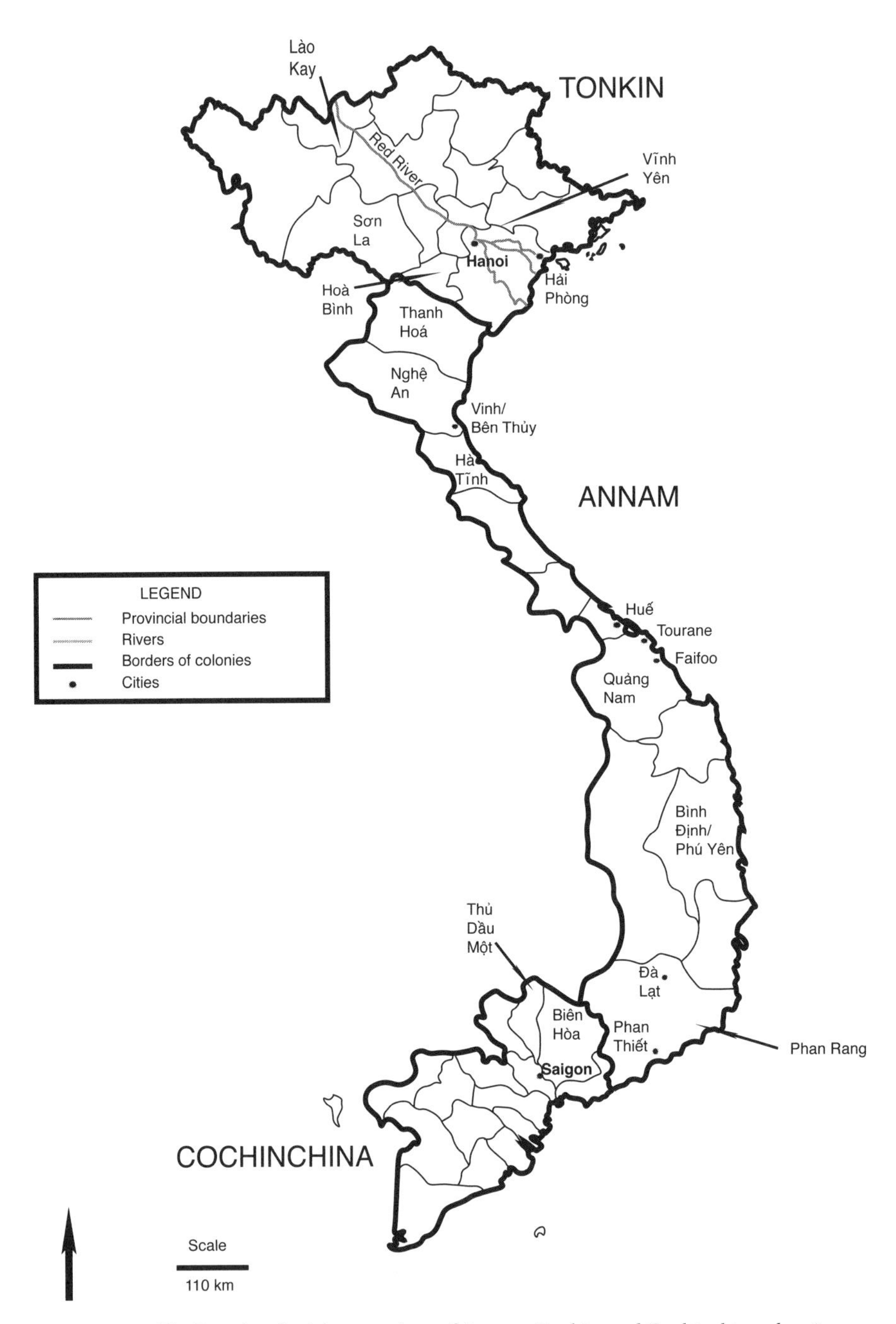

MAP 1.01 The French colonial possessions of Annam, Tonkin, and Cochinchina, showing provinces and cities mentioned in text. Base map from 1914. Brenier, *Essai d'Atlas statistique de l'Indochine française*. Redrawn by author.

The Indochinese laws were considered to be so well constructed and comprehensive that they were held up as an example for laws in other colonies, such as Madagascar and West Africa.[9]

However, despite the comprehensive laws, the institutions that were needed to enforce these regulations emerged slowly. The new Forest Services began to hire personnel but had to be led by French men (the heads of departments were additionally required to have attended forestry school in Nancy, France), thereby making it difficult to recruit (Thomas 2000).[10] We can see the challenges by looking specifically at Annam, the site of the most extensive forest resources within Vietnam. In 1906, a Forest Service management structure arose, which divided the protectorate's forests into individual cantonments (*cantonnements forestiers*). Each cantonment was to be exclusively run by a European agent, under whom were forest divisions and posts staffed with patrol guards, who tended to be locals.[11] By 1919, Annam claimed six cantonments, headquartered in Bến Thủy, Thanh Hóa, Huế, Tourane, Phan Thiết, and Đà Lạt cities, employing 30 European agents and 156 native forest guards. However, ten years later, only 42 European agents and 232 guards served for all the forests of Annam, an area of approximately six million hectares (Fangeaux 1931).[12]

FOREST PROBLEMATIZATION AND THE EMERGENCE OF ENVIRONMENTAL RULE

The creation of a Forest Service required the identification, mapping, and classification of forest stands, to which the new regulations on use and exploitation would be applied. The laws on forestry for Indochina had been passed before any specific scientific assessments of the type of forests and the requisite silvicultural treatments that might be needed to produce timber sustainably had been undertaken; forests were simply created through legislation. Officials made initial determinations of the ecological types of forests visually, without systematic data collection, and these proved deceptive. Labels of "*forêt inexplorée*" in maps of the time promised riches once these areas could be explored. Yet once inventories were undertaken of forest plots, these seemingly abundant forests revealed a worrying problem: high species diversity and low numbers of high-value timber trees. Part of the diversity was attributed to the geography of the colony; one could find tropical timbers that were almost "equatorial" in the South, while in the North, forests were closer to the Mediterranean timbers of France (Mangin 1933, 42).

The heterogeneity was remarkable. In just four hectares surveyed at the turn of the century in Biên Hòa, Cochinchina, foresters found 1,080 trees greater than 10 cm diameter, but 80 percent were considered "useless varieties"; only 266 specimens had an economic value, and in this plot only nine trees had a diameter large enough for commercial exploitation (Gourou 1940, 505). The head of the Indochinese Forest Service put this diversity in perspective in a comparison with France, where there were approximately 100 kinds of trees, of which 80 had uses for forestry, whereas in Indochina, he estimated 1,200 tree species, of which only 91 were useful for forestry, or "a mere 8 percent" (Mangin 1933, 648). This realization was a rude shock to the nascent Forest Services—unlike in Java or British India, there were no extensive stands of dominant and valuable timbers like teak (*Tectona grandis*) or sal (*Shorea robusta*) that could be clear-felled.[13] Adding to the problem, physiological studies showed that the Indochinese forests were not just poor in value, but slow growing:

> Good species grow much slower it seems than should happen in a tropical climate. It seems that hardwoods increase by no more than a centimeter a year in diameter, suffering from exposure for many years from dominant softwoods in the vicinity which grow more rapidly, and thus hardwoods require many years to become large logs. In addition, this slow development makes it difficult, without human intervention, to undertake the improvement of impoverished forests. (Service scientifique de l'Agence économique de l'Indochine 1931, 1–2)

It was estimated that vast areas of the Indochinese forests would produce far less than 100 cubic meters of usable timber per hectare (and many areas perhaps only 10 to 20 cubic meters), against 200 cubic meters or more for temperate forests back in Europe (Thomas 1998). Of nearly 37 million hectares of forest land in the Indochinese Union that were designated by the Forest Service in a 1925 mapping exercise, only 6 million hectares were considered "rich" forests—capable of producing more than 300 cubic meters of timber per hectare (Mangin 1933). Further, the topography of Annam and Tonkin in particular made industrial scale extraction even more difficult. As the head of the Annam Forest Service complained:

> many areas are inaccessible, due to lack of road penetration, and the only forests we can really exploit are those that are penetrated by deep watercourses which can be used to float out the timber. It follows that

> the richest forests cannot be exploited, while forests located in lowlands or on the foothills of the Annamite chain are already partially exhausted. (Fangeaux 1931, 230)

The open forests of the central plateau of Annam were considered better for exploitation than the high mountains in Tonkin and along the Annamese mountain chain, but in those areas the Forest Service was often under competition with agricultural concessions for the best basaltic lands (Cleary 2005b). In still other areas, such as coastal mangroves or swampy inland cajuput forests (*Melaleuca leucadendra*), trees were difficult to access and did not produce wanted timber. As a result, these areas often fell under the Public Works Department as wastelands (Biggs 2005).

Foresters soon identified a clear reason for both the heterogeneity and the lack of dominant timber trees. Contrary to appearances, most of Indochina did not consist of primary or virgin forests but rather secondary regrowth modified by human impacts: "Woodland and bushland can be natural (due to bad growing conditions, poverty and lack of soil depth); but most often, their origin is artificial and represents a stage of deforestation standing between poor forest and grassy forest; in this case they are due mainly to bush fires and indigenous clearing" (Mangin 1933, 646). The heterogeneity of Indochina's forests was not seen as something "natural" or a function of ecological or biophysical conditions: it was human-caused. As such, officials searched for people to blame for this state of affairs. As geographer Pierre Gourou emphasized to authorities in the late 1930s,"it is imperative to make it clear that 'primeval' forests, that is, those that have never been exploited or modified by man, are rare. Most of the forests have been exploited by man, who has extracted the most precious varieties from them, or else they have been ravaged by *rẫy*" (Gourou 1940, 462–3).

The practice of *rẫy*, the Vietnamese word for swidden or shifting cultivation, whereby a forest area is burned and planted for annual crops, emerged as the avowed enemy of the Forest Service. *Rẫy* became one of the few Vietnamese words to be adopted in common parlance among foresters, who had no French equivalent, thereby stressing its foreignness.[14] In a classic example of identifying a problem for which a ready-made solution could be applied, French foresters chose to identify swiddening natives as the issue in need of improvement, rather than the potentially more intransigent problem of an entirely unchangeable climate and ecology defeating their plans for high-value timber logging. In other words, the problem was not the trees: it was the people. In a document prepared for the 1931 Colonial Exhibition in Paris, colonial officials bemoaned the sad state of native management of forests:

> Depletion is the rule wherever the indigenous people abuse the forest. In areas that are not monitored, the mountain tribes do not concern themselves to ever replace the material they remove. Free cutting is destructive of valuable species. In addition, fires, usually due to carelessness of the inhabitants, annually devastate vast areas. . . . Finally, the methods of cultivation of the mountain tribes . . . quickly leads to real destruction of the wooded area. Lacking irrigated rice and gently sloping fields, they usually clear by fire those areas of forest favorable for cultivation, and amongst half-burned trunks, they grow rice, maize and cassava for three to four years, and then when the good earth surface which has been enriched by ash was swept away by runoff waters, they abandon the land and move away. This is the *rẫy*. Once the field is exhausted, the forest soon regains its place, but good species are replaced by worthless woods, bamboo, lianas, or wild bananas. Huge areas are ruined piece by piece, and we do not expect this situation will improve, as it has not been possible to settle these wandering tribes. (Service scientifique de l'Agence économique de l'Indochine 1931, 2)

The emergence of environmental rule was thus set in motion. While officials in fact minded the lack of "monitoring" of indigenous inhabitants, their "carelessness" and mobility, not to mention the fact that swidden fields competed with the Forest Service for lands to produce economically valuable timber, they emphasized the ecological "damage" of swidden, and used this concern to identify interventions that needed to be made.

INTERVENTIONS IN THE COLONIAL FOREST

The perceived solution to the dual problems of wanton native exploitation that could not be monitored and the lack of economically valuable dense timber stands was to create new categories of forest for which uses would be restricted and management guided by the Forest Service. Starting around World War I, the Forest Service classified the forest estate into the Reserved Forest Estate (*domaine forestier réservé*), requiring specific management plans for its harvesting, and the Protected Forest Estate (*domaine forestier protégé*), where forest exploitation and produce were taxed but where specific management plans were not in place.[15] A 1913 Forest Law noted that reserves "will be created, delineated, surveyed and registered to exclude free cutting and manage rational exploitation. These operations are designed to protect and improve the water regime and ensure orderly production of wood" (Sarraut 1913, 233). The Forest Law also established restrictions on harvesting certain valuable species of classified timber (such as *Erythrophleum fordii*, known as *lim*) found outside reserves (Guibier 1918).

For the Protected Forests category, management was minimal. This category included nearly all areas not designated as Reserved Forests, with the exception of those that were privately owned, where regulations did not apply. In so-called "Protected Forests," wood extraction was primarily subject to regulation and taxation, where an officer from Forestry, Customs, or the indigenous guard checked harvested wood and assessed fees on the type of wood removed. Logs for which taxes had been paid were to be notched with a distinctive mark and to have paperwork indicating the name of the feller, tax amounts paid, and territory the wood was to be transported to; however, these rules were "usually ignored when the cutting site is too far from a ranger station," the head of Annam's Forest Service admitted (Fangeaux 1931, 233). The imposition of new permits and taxes, even on low-value firewood from what were previously community-used woodlands, raised numerous local protests, but forest officials responded that these costs were necessary to encourage users to follow instructions on minimum felling diameters and limitations on the amounts of wood allowed for removal.

The so-called "Reserved Forests" received the bulk of attention from the Forest Service on how to exploit and reforest them effectively. Despite the name, these areas were not selected based on ecological criteria and cannot be considered "conservation" areas. Instead, these reserves were inaugurated mostly near "major communication routes (rivers, seashores, roads, canals, railroads) as close as possible to a place of labor" (Baillaud 1914, 206) so that they could be harvested more effectively (see figure 1.01). In areas targeted as possible reserves, officials initiated a commission for approval, composed of the administrator of the province, the chief of the Forest Service, and the heads of the cantonments and divisions to oversee the forest reserve, along with local chiefs. One month before the meeting of the commission, postings and signs along the perimeter of the new reserve brought the proposed borders to the attention of interested parties (Cleary 2005a). The commission was to assess the issue through at least one joint meeting and send its report to the governor-general and seek the opinion of the Chamber of Agriculture

FIGURE 1.01 Logs of *Erythrophleum fordii* (Vietnamese: lim) are floated from upland areas of Hà Tĩnh and Nghệ An down to the port of Bên Thủy in Vinh city. Photograph from *L'Indochine Française*, Publications du Gouverment Générale de la Indochine, 1919.

as well. The reserve would then be announced in the official journal and the provincial newspaper, as well as being

> brought to the attention of all concerned villages and villages through posters in community houses. Reserve plans, drawn up by the Forest Service, shall be deposited in the chief town of the province and . . . the head of each township and village positions in or near the reserve. Plans submitted to the villages indicate the full extent of the reserve to be located on their territory. The reserve is finalized upon posting of these orders. (Sarraut 1913, 233–4)

After a reserve was created, it then became illegal to go into it except by public roads. Not surprisingly, this formalized process did not provide local communities that were to be affected with much say over reserve boundaries.[16] To add insult to injury, local villages were often required to provide the corvée labor needed for building forest roads to open the reserve for exploitation—the very roads that would now be forbidden to them.[17]

To properly manage these Reserved Forests for a steady supply of wood in perpetuity, primarily through rotational cutting as practiced in the metropole, the areas had to be carefully demarcated and then surveyed meticulously to ensure timber production was subject to "methodological exploitation."[18] Foresters undertook experiments in cutting and replanting, such as the selection felling and coppice with standards methods that were the gold standard of Nancy training. As authorities noted,

> The set-aside [reserve itself] is not sufficient to ensure rapid improvement of the forest: it must be complemented by the development of in-depth study of the material it contains, finding the best treatment methods to be adopted, regeneration and propagation of the best species or those that best suit the needs of the region for food and farming operations, the improvement of cuttings or extraction, protection against fire, etc. (Service scientifique de l'Agence économique de l'Indochine 1931, 7)

Each Reserved Forest was to be mapped and visualized; it was to have a clear working plan with an appropriate cycle of felling (*coupe méthodique*) which would enable an eventual return to site of the first harvest. One forester noted with satisfaction the 250-year rotations of oak that had been achieved under Louis XIV, and expressed similar hopes for Vietnam (Kernan 1968). These working plans literally created the basis for the valuation of "forests" for the colonial administration; a true scientifically Reserved Forest did not exist in the minds of foresters if it was not bounded by a working plan.[19]

Yet rotational coupe logging worked poorly in Indochina because large stands of dominant species that could be block felled were hard to find, which made it nearly impossible to calculate how the hundreds of tree species in diverse stands would regrow after logging. The French soon found that ecological conditions pushed back against their master vision for "scientific" working plans:

> The exploitation of Indochinese forests therefore presents difficult problems that the European forester is not prepared to solve; in Europe it is sufficient to establish a methodical rotation of cuttings in order to assure a rational exploitation and the regeneration of the forest. We have seen that in Indochina the most methodical cuttings infallibly lead to an exhaustion of the forest without hope of regeneration. (Gourou 1940, 508)

After logging, spindly forests of softwood replaced the former hardwood dominated stands in reserves, frustrating colonial foresters to no end. For example, in one area in the Red River Delta, 71 percent of surveyed trees in 1929 in the Nang Yên forest reserve were smaller than 20 cm in diameter after logging, meaning that the forest had essentially been reduced to "a thicket, in short" (Thomas 2009, 114).

Despite these failures, the number of Reserved Forests steadily grew. According to the Forest Service, more than half of the total land area of Indochina was composed of forest, and between 1891 and 1902, some fifty reserves were created, mainly in Cochinchina. By 1912, 220 reserves had appeared, with a total area of 490,000 hectares (Baillaud 1914). In 1919, Annam alone had 56 total reserves, totaling 397,291 hectares, despite having only a handful of people to oversee them.[20] The visualization and mapping of new reserves continued apace. In 1925, a new 1:100,000 scale map of the forest resources of the colonies was drawn up by the Forest Service, the Public Works Department, and the Geography Service, which, for the first time, used aerial photography to verify forest stands and attempted to provide a country-wide visual assessment of forest resources.[21] Authorities classified six types of land cover on these maps, based on an explanation provided by Indochinese Forest Service Chief Mangin, who oversaw these efforts:

> I resolutely eliminated from our vocabulary terms such as primary forest, secondary forest, xerophylle forest, tropophylle forest, open forest, flooded forest, Dipterocarp forest, Euphorbiaceae forest, etc., as too general or too specialized, on whose value we are far from agreeing. I established three simple types, easily discernible and comparable, based on the consistency of forest stands and, for each type, one or more value classes according to their composition. (Mangin 1933, 644-45)

The top classification was "rich forest" that produced 300m^3 timber of high quality species per hectare (an estimated 8 percent of the total Indochinese forest estate). Medium forest was classified as that which produced 150–300m^3 timber per hectare (33 percent of forests), poor forest supplied only 100–150m^3 timber with low numbers of choice species (10 percent of forests), and the rest of the forest estate was woodland and open areas that supplied no valuable timber at all (Mangin 1933). As this classification system shows, rather than merely being "simple" and "comparable," sorting forests this way used primarily economic criteria, not ecological ones, as a prime distinguishing factor.

By 1931, Annam had 112 Reserved Forests with a total surface area of 854,593 hectares and nearly 500 Reserved Forests extended across the rest of the Indochinese Union (see map 1.02).[22] In Cochinchina, nearly 36 percent of forests were "reserved," while the figure was 14 percent and 10 percent for Annam and Tonkin's forests respectively. However, many of these reserves "existed mostly on paper" (Thomas 2009, 112). The estimated number of surveyed Reserved Forests with an actual working plan was highest in Cochinchina, with nearly half being actively managed, to a low of only 6 percent of Reserved Forests in Annam (see table 1.01). Thus, the "scientifically managed" forest estate of Indochina, while seemingly extensive on paper, actually accounted for only 1 to 2 percent of total forests, with the exception of Cochinchina, where about 15 percent was under active management. Perhaps even more surprisingly, there was also a complete absence of protected areas for stricter conservation of flora or fauna in Indochina, even though they existed in other French colonies by the 1930s, particularly in Algeria and Madagascar (Ford 2004). Clearly, in Indochina, economically valuable timber, not ecologically important species or landscapes, drove Forest Service attention.

As a result, the Reserved Forests were not spread evenly throughout the colonies. Rather, they tended to concentrate in logging-accessible areas with high-value timber. Within Annam, the number of forest reserves per province ranged from a high of twelve in Thanh Hoá to none at all in some areas; the province with the most land area under Reserved Forests status was Hà Tĩnh, with 124,000 hectares (see table 1.02). The Central Highlands, the interior of Annam bordering Laos and Cambodia, had no Reserved Forests, despite this being a heavily forested area, because these lands had not been "pacified" by military means yet, and there were no rivers that ran down to the coast to facilitate the removal of timber. This unevenness meant some local communities faced harsher restrictions on forest use than others, and it is perhaps no surprise that Hà Tĩnh, with the largest areas of Reserved Forest of any province, later became the epicenter of protests against colonial rule in 1930 (see map 1.03).

Given contested regulations and the low numbers of forest guards, most cut wood in Indochina continued to be supplied from the less regulated Protected Forests, not from Reserved Forests, and most of these areas failed to meet even their modest goals of a sustainable timber supply (see table 1.03). Competition with mushrooming agricultural concessions additionally resulted in some declassifications of once-Reserved Forests in favor of the expansion of tea, rubber, and coffee plantations in the interwar period.[23] Concerns about deforestation in these non-Reserved Forests continued to grow; officials estimated around 12 million hectares of impoverished and

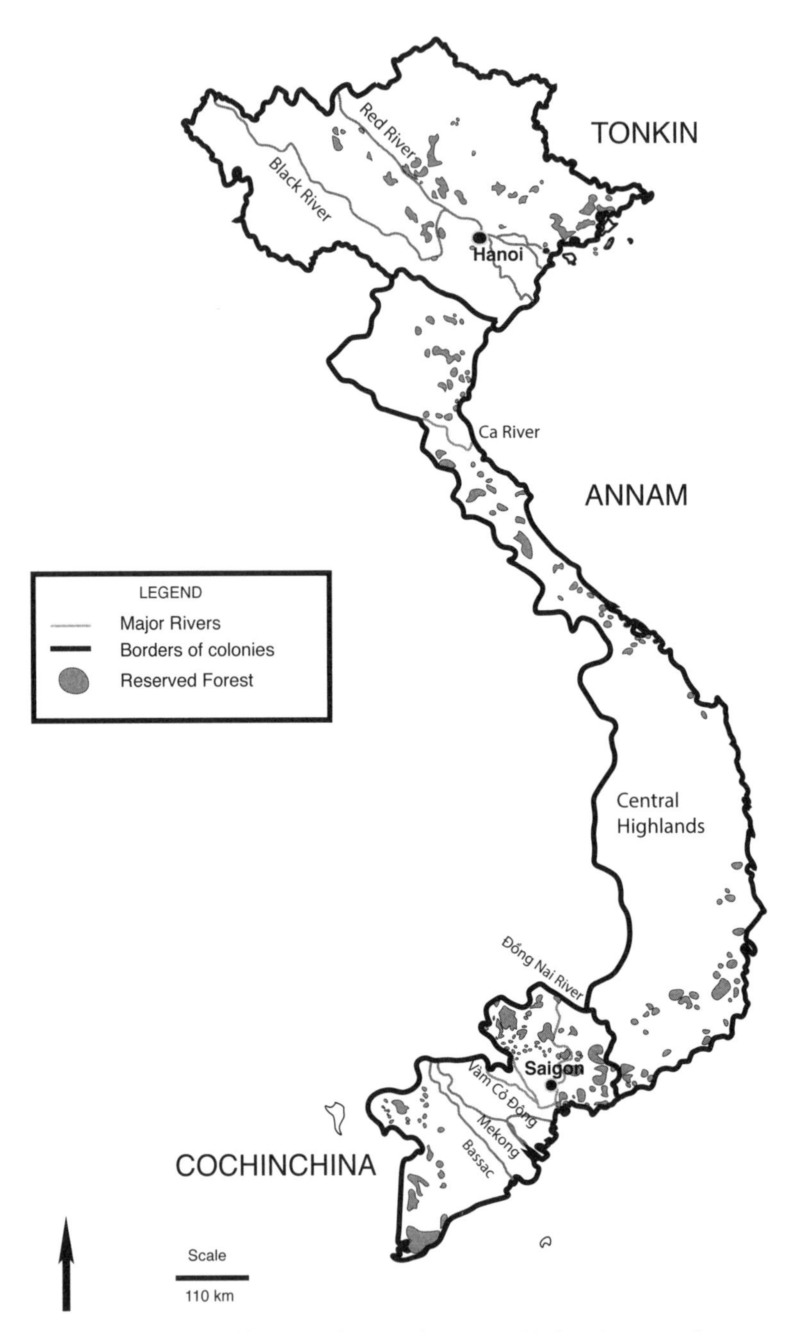

MAP 1.02 Extent and location of Reserved Forests in Tonkin, Annam and Cochinchina, circa 1940. Base map from Maurand 1943. Redrawn by author.

TABLE 1.01 Forest estate, forest reserves, and actually managed forests in 1930

Colony	Total Forest Estate (ha)	Total ha of Reserved Forests	Estimated Percentage of Forests that were Reserved	Estimated Area of Reserved Forests Actually Managed with a Working Plan	Percentage of Reserved Forests with Direct Management	Percentage of Total Forest Estate with Direct Management
Cochinchina (total land area: 5,500,000 ha)	1,811,111 (27% forest)	652,000	36	278,000	42	15
Annam (total land area: 15,000,000 ha)	6,271,428 (42% forest)	878,000	14	56,000	6	Less than 1
Tonkin (total land area: 10,500,000 ha)	3,450,000 (30% forest)	345,000	10	54,000	16	1.5

Sources: Data from Maurand 1943; Service scientifique de l'Agence économique de l'Indochine 1931; Thomas 1999.

TABLE 1.02 Geographical distribution of forest reserves in Annam, circa 1925

Province	Total Number Reserved Forests	Total Area Reserved Forest
Hà Tĩnh	7	124,000 ha
Phan Rang	5	116,800 ha
Thanh Hoá	12	95,500 ha
Nghệ An	10	77,010 ha
Thừa Thiên Huế	9	27,924 ha
Quảng Trị	3	20,325 ha
Quảng Nam	7	16,027 ha
Quảng Ngãi	1	3,500 ha
Phú Yên	0	0
Bình Định	0	0

Source: Own calculation from *Carte economique de l'Annam* Service géographique de l'Indochine, 1925. Map held by Library of Congress, Geography & Map Reading Room.

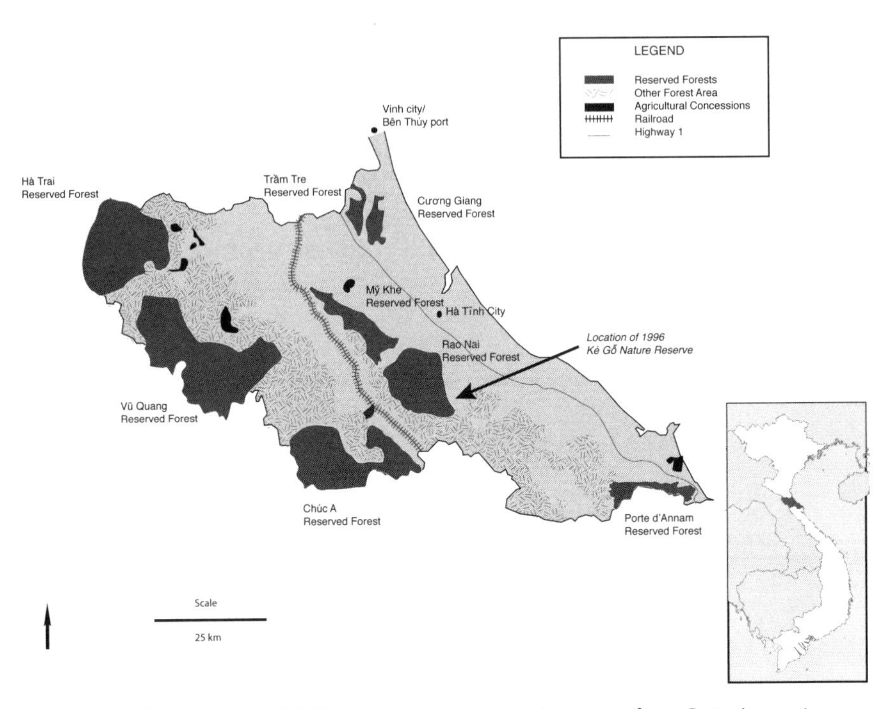

MAP 1.03 Forest extent in Hà Tĩnh province, circa 1925. Base map from *Carte économique de l'Annam, Province de Ha Tinh*, Service géographique de l'Indochine, 1925, held by Library of Congress Geography and Map Reading Room. Redrawn by author.

depleted woods by 1933, with Annam and Cambodia perceived to have the fastest rates of deforestation (Mangin 1933). Consequently, by the 1930 revisions of the Forest Law, the reservation system was essentially abandoned, and all forests identified as either classified (*domaine classé*) (encompassing all previous Reserved Forests, communal forests, and reforestation areas) or protected (*domaine protégé*) (all other forests, with nearly all non-state uses restricted) in an attempt to reduce excessive deforestation rates. While this more restrictive system garnered 3,500 official protests and counterproposals during the drafting period, the new law came into force for all divisions of Indochina by June 1931 (Mangin 1933).

PRODUCING NEW JUSTIFICATIONS, ENROLLING NEW ACTORS

Given the multiple challenges the Forest Service faced, by the 1910s and 1920s, officials began searching for other justifications for the continued reservation of forests other than economic profitability and sustainable timber production, and it is at this time that we see the extension of environmental

TABLE 1.03 Forest product production in Indochina in 1929

	From Reserved Forest	From Protected Forest	Total
Lumber	24,747 m^3	928,137 m^3	952,611 m^3
Firewood (stere)	447,377	1,365,475	1,812,852
Charcoal (tonnes)	3,146	15,190	18,336

Source: Service scientifique de l'Agence économique de l'Indochine 1931, p. 8.

rule to new audiences and areas. In particular, the role of forests in regulating water and climate became increasingly emphasized in reports from the Forest Service. Given the importance attributed to forests in preventing landslides and floods in France in the 1800s, Nancy forest school graduates would have been well aware of the arguments for the role of forests in preventing hydrological changes (Cleary 2005a).[24] Theories about forests as "large water tanks that act as sponges" were widespread at the time, and forest officials in Indochina increasingly marshaled data and evidence to bolster these claims, such as experiments that purported to show that the "absorption of rainwater by the ground is 6% in the bare land against 61% in woodland" (Bertin 1924).[25]

Forest officials saw opportunities in presenting this modern "scientific" evidence about forestry to new audiences in journals such as the *Bulletin of the Friends of Old Hue* and the *Economic Bulletin of Indochina*. In this way they acted to enroll other audiences as advocates for forest reserves, engaging in the process of *interessement* required to build a network for environmental rule. A clear example of this can be seen in a series of articles that Roger Ducamp, the first head of the Indochinese Forest Service and a top graduate from Nancy, wrote for a non-forestry audience in 1912 regarding the problem of siltation in the economically important harbor of Haiphong in Tonkin. Ducamp confidently argued:

> the question that arises here in Haiphong is fully linked to the 'Forest Plan.' One cannot study thoroughly and seriously the 'Water Plan' without looking into the imperative of forest issues and without reference to research that raises the solution to the problem of conservation of forests. . . . Any comprehensive study of the water system needs a preliminary study of the forests on which they depend. . . . I will try to show that it is easy, if you wish it, not only to keep the port of Haiphong open to navigation without artificial dredging, but it is also possible to

> return navigability of large ships to the Red River up to the Doumer Bridge, which would make it a worthy brother to the Saigon River. To achieve this, what can we do? The argument that I want to make obliges me to recall the importance of the role of the trees and forest groves in the important questions of navigability and water management. (Ducamp 1912a, 494)

Ducamp argued that costs of new roads and dredging for Haiphong could be more profitably spent on forest conservation in the highlands of the Red River watershed (he suggested 700,000 hectares of reserves, nearly three times the existing number), which, he added, would also serve the purpose of supplying wood for shipping barrels needed by Haiphong industry. Other articles by Ducamp in different venues stressed the importance of forests in regulating microclimates, such as transpiration and dew points, contending this would help agricultural development if forest cover was retained in surrounding areas (Ducamp 1912b). Through these arguments, Ducamp and other officials attempted to persuade citizens that forests should be protected: for military reasons, economic reasons, and aesthetic reasons—each audience had its own justifications.

Huge floods in the lower Red River in 1915, which nearly submerged Hanoi, the seat of colonial government, provided foresters with further claims for the need for forest reserves. Extended rains in July led to the rapid rise of the river system, breaking the protective river dikes in nearly fifty different places, resulting in hundreds of deaths and the inundation of 100,000 hectares of rice land in the lower delta (Hoàng Cao Khải 1915). Engineers and hydrologists focused on an aging levee system and poorly maintained dikes as the primary reason for the disaster (Peytavin 1916), but foresters also joined the discussion, pointing to the "dizzying speed of deforestation" in Yunnan and Upper Tonkin as the culprit that would "ruin the Delta" (Sion 1920, 316–17). In subsequent requests for forest reserves, particularly in mountainous areas, local forest officials often used water catchment and flood control as justifications for their entreaties. Visuals indicating worries about "bare hills" (*collines chauves*) and "gullies" (*ravinements*) that resulted from deforestation and affected hydrology circulated in articles and postcards (see figure 1.02). As forest service officials relayed:

> As in any part of Annam, the results are convincing: forest loss that is growing without anything to replace it, gradually increasing the extent of wastelands at the foot of the mountains, because deforestation is followed immediately by gullies; landslides of riverbanks after the destruction of

FIGURE 1.02 French aerial photo of bare hills (*collines chauves*) in Tonkin circa 1920s. Photo by Aviation Militaire Indochine.

> trees or shrubs that held back these banks. . . . In the past, huge areas were covered with intact pine forests, rising out of the mountains as thick bouquets on the peaks and the humid valleys; now they are barren hills trampled by herds; rains beat the land; and all the hills deforested by fire between Huế and Thanh Hóa are currently furrowed by ravines and end up unable to even feed livestock. (Guibier 1918, 34)

These professed concerns regarding the environmental services provided by forests perhaps have the ring of early conservationist impulses. But while forest officials may have used ecological concepts to justify the protection of some forests, in reality, even the forest reserves cast in hydrological terms were primarily about economic exploitation, according to historian Frédéric Thomas (Thomas 1998). Thomas's thesis goes some way in explaining why, if the French authorities were so concerned about the deforestation and resulting downstream changes, they did not institute stricter reserves in which all exploitation (not just local use, but all logging entirely) would be prohibited. Such nature reserves existed in other French colonies—Madagascar, for example, boasted at least twelve national parks—and the Dutch East Indies had also instituted strict protection for some forests at this time (Kies 1936). There had been requests from botanists and other scientists to enact these in

Indochina as well, but these calls for conservation reserves went unheeded, and much of the state forests continued to be targeted for logging.[26]

PROBLEMATIZING AND CLASSIFYING FOREST SUBJECTS

In addition to the importance paid to the climate regulating effects of forests, rising attention was also paid to the ethnic dimensions of local resource use, and these two projects were clearly linked. From the earliest days of forest regulations, officials derided local use as chaotic and wasteful, whether fuelwood extraction or clearance for pasturing, and colonial officials held both Vietnamese (known as *Kinh*) and the ethnic minorities who used forests in poor regard. But increasingly attention turned to the mountainous minorities as the primary culprit in forest loss.

There are likely several contributing factors to this shift in discourse. First, the Forest Service predominantly reserved and managed the more accessible forests in the lowlands, and subjected the activities of Vietnamese to permits or working plans. This left only mountainous forest outside the reach of the Forest Service, and new justifications were needed for the expansion of activities there. Secondly, colonial agriculture, particularly rubber plantations, rapidly expanded after World War I, and often competed with indigenous minorities for land. Plantation owners vociferously complained about the destructive impacts of native fire on new rubber plantings and urged restrictions on this practice.[27] Finally, there was increasing attention to ideas of "native improvement," particularly after World War I, with arguments in favor of colonialism as a "civilizing mission," rather than simply an economic one, prominent at this time (Conklin 1997). Colonial officials in Indochina professed interest in helping native peoples achieve higher economic status through interventions in science, technology, and medicine, and these policies for native improvement and development (*mise en valeur*) included getting minorities to practice what were seen as more civilized and profitable forms of agriculture (Aso 2011). Happily for officials, these concerns over the cultural and economic differences of minorities could be folded into policies on forest management through interventions of environmental rule.

Whereas officials had begrudgingly tolerated swidden in the early days of the colonial enterprise (Cleary 2005a), by Cochinchina's 1912 Forest Law, swidden could only be done "by special authorization only in areas determined in advance by the Administrator and chief of the province, in agreement with the Forest Service" (Sarraut 1913, 238), and by the 1914 Forest Law revisions, swidden was prohibited entirely (Thomas 1998). The arguments against ethnic minorities' use of swidden took many forms:

economic, cultural, and environmental. Administrators emphasized the unruly, even "barbarous" nature of these activities, contrasted with the more modern and rational Forest Service practices. Swidden was considered autarkic as well, focused only on subsistence production, which prevented minorities from engaging in more civilized market activities. There were even references to swidden fitting with the "natural indolence" of mountain dwellers, since they "hate the work of rice planting, do not like plowing, and find it much easier to burn the forest, not worrying about what the results of this vandalism are."[28]

But it was the environmental impacts of swidden that began to dominate discussions, particularly after the 1915 floods. As the Résident of Sơn La Province noted in a letter to his Hanoi superiors:

> I think it would be appropriate to put an end to this deplorable practice of *rẫy* if we are to prevent complete deforestation of the Mountainous region [sentence underlined] which cannot be delayed. The worst, in my opinion, is that the destruction of forests has very dangerous implications for water resources in the Delta and floods would certainly be much less serious and disastrous if the upper region of Tonkin was still forested [underlined]. It seems not too late to try to fix this sad state of affairs, and section 183 of the Penal Code provides enough weapons.[29]

Forest officials also began to try to enroll native authorities, such as Vietnamese mandarins, in their fight against swidden, arguing it was in their own interest to avoid downstream flooding; for example, the Résident of Hòa Bình province asserted that:

> I have warned the native authorities against the disadvantages that result from irrational deforestation on the side of the mountains themselves, the '*rẫy*' system, which destroys everything in bushfires, and which stops the growth of young shoots and prevents reforestation. The mandarins are perfectly aware of the danger that they would face in their lives and their property among the people of the plain, by floods and avalanches, which would be made possible by the loss of [forest] roots, which have enclosed the land of their ancestors and maintained its compactness.[30]

Yet these concerns over swidden and the protection of water supplies hid other economic reasons, as swidden competed with the production of lumber for railways, other public works, and mines. If foresters were truly concerned with conservation of environmental services and hydrology, they would have banned all forest use, including state logging, in highland areas, not just the practice of swidden.[31] Further, despite the voluminous studies carried out on forest inventories, test plots for selective felling, and reforestation in Reserved Forests, there appear to have been no studies of forest growth after swidden practices to verify if the dire predictions of decreased biodiversity and softwood dominance after forest burning or clearing were actually true. In this, environmental rule privileged knowledge that fit state interests (continued logging in Forest Service–managed forests) and ignored situations where additional or new knowledge might have contradicted predetermined interventions (stopping swidden).

Interventions against swidden took many forms, ranging from mild requests from authorities to restrict these practices to certain times or locales, to more serious actions against what one official called "the bankruptcy of the woods" (Guibier 1918). Some officials, particularly in the north of Tonkin where swidden was practically the only way of producing food crops on sloping lands, sympathized with swiddeners and appeared to take few actions against it.[32] Others argued that measures against swidden needed to be multifaceted, including education, supervision, and punishment (Allonard 1937), and in areas where swidden was widespread, local commissions cropped up, composed of the head of the province, the head of the forest division, and several local authorities.[33] These commissions discouraged the use of fires to clear fields by holding entire villages responsible if any forest damage was reported from their area, as well as punishing individual violators, such as banning them from raising livestock for five years (Thomas 2000) or requiring them to reforest any areas that had been set on fire due to "malice" or "self-interest" (Guibier 1918). This system of collective punishment was relatively successful, as the Forest Service of Annam reported in 1919:

> Many fires continue to be reported each year and it is not possible to catch an offender in the act. In most cases, however, if you do not know the offender, at least we know for sure what the villages they depend on are, and the Administration should be able to take action against these villages. . . . [In] the reserve of Hon Lu, a fire broke out where tree seedlings were being planted. After that it was made clear to the people that this disaster was likely due to one of them, and they agreed to provide the labor needed to reforest the burned part. Last year, four hundred offenders were already available to the Chief of Division to reforest burned land near their villages. If these facts were repeated more often and were brought to

> the attention of the people, perhaps it would be possible to stop this fatal problem.[34]

An additional proposed solution was to provide highlanders with agricultural tools of the lowlands, in the hopes that they would take up wet-rice cultivation and abandon their culture of swidden; some even argued in favor of bringing in Vietnamese, Thai, and Chinese to demonstrate terraced-rice agriculture in the mountains, noting that "without doubt it is always difficult to change this way of life, but a transition to sedentary culture can be observed in several parts of India, in Burma, and in our colony, and since prosperous alluvial plains are a vital part of our possession, it is urgent to put an end to destructive primitive nomadism" (Sion 1920, 317). In a few other areas, authorities resettled minorities away from forested slopes into valleys, stressing not only the environmental benefits of such actions, but the fact that resettlement could change the culture and identity of highlanders as well:

> The attempts in this direction . . . are very encouraging. Several Moi [*minority*] villages have asked to follow the example of the resettled village of Rong Hong, now installed for over a year [in a valley]. . . . All efforts are made to bring the Moi to change the way of their culture. Sometimes we try to fix them in the plain, sometimes they are used for the execution of forestry work, as we believe that every Moi taken from the mountain represents one *rẫy* less. Unfortunately, there are still very few who are willing to stay in the plains, many have tried, and then suddenly disappeared, taken by the yearning for solitude and heights.[35]

Overall, however, the lack of success of most resettlements, and dearth of money in the Forest Service for these activities, prevented much extensive action on sedentarization. Pointedly, solutions that linked agriculture and forestry, such as the use of the *taungya* system in Burma, where local minority cultivators such as the Karen planted and cared for commercial timber like teak in their shifting cultivation fields, were not replicated in French Indochina (Bryant 1994). Timber production was for the state alone to manage, and any suggestions to the contrary were not on the table.

COUNTER-CONDUCT IN THE COLONIAL FOREST

There were several consequences of the increased attention to ethnic minorities' forest use. Not surprisingly, many objected to the stricter restrictions on swidden and fire, and engaged in passive resistance by treating state-managed forests as a continuing source of local livelihoods. But perhaps even more significantly, the attention to minorities' swidden practices may have blinded the Forest Service to the increasing complaints from ethnic Vietnamese about higher taxes and new charges that were applied to them. These complaints were particularly strong in the northern part of Annam, which had the most extensive forest reserves. While changing subjectivities are difficult to assess on the basis of archival evidence alone, patterns of resistance to colonial forest restrictions provide indications of how the new environmental rules led to counter-conduct in surprising ways.

Resistance to the new forest controls took multiple forms. As noted earlier, many villages did not understand what forest reservation was, and, when invited to meetings, simply refused to show up; however, this was taken by colonial authorities as a sign of acquiescence, not resistance. Even when villages did actively protest, their concerns often fell on deaf ears. For example, the Résident of Vĩnh Yên wrote to Hanoi in 1914 after a new forest reserve was declared:

> I am pleased to announce the results of the survey conducted among the natives ... regarding their rights that have been reserved by order of last July 8. The natives questioned on this subject are silent, but through their municipal authorities they issued the following statement, "We have no knowledge of the order of July 8 last" [and] "Our circumstances have been this way for a long time, and we plant our crops on the plains and in the mountains in areas that are already cleared. As for cutting wood, bamboo or making thatch huts that we need, whenever we ask the Forest Service, they make it very difficult in giving us permission." When encountering these conditions, I felt obliged to show the interested municipalities by affixing on the *Đình* [local communal house] a copy of their use rights [for forests], with a translation into both *Quốc Ngữ* and characters, as defined in the order of the Governor-General on June 28, 1909.[36]

Simply showing the Forest Law to local peoples in the form of an official handbill nailed on the wall of the communal house seemed a clear enough statement of authority to this official, who did not seem to understand why subjects might not want to conform.

Not surprisingly, villagers had little recourse but to continue to use their forests, but with the added inconvenience of trying to avoid forest service guards, or using petitions to access what had once been theirs, and the archives burst with letters asking for use rights for nearby forests. An interrogation of the head of a village with a high number of "forest violators" by an official from the Résident Supérieur of Tonkin's office shows resistance to the new subject formation that forest laws required took many forms:

> Q (Representing Résident Supérieur of Tonkin): Do you know that the government has given permission to landowners in your commune to clear a piece of rice land in the forest equivalent to four western hectares to expand their fields, and after expanding your rice field you have permission to harvest wood without taxes in that parcel that is allotted to you, so long as the wood is for fuelwood and housing only, as well as thatching and other non-timber forest products that are needed in the forest, but you must exploit according to the permission granted: do you understand this?
>
> A: No.
>
> Q: Of the landowners in your village who have rice fields near the forest, is there anyone who has cut down trees near the rice fields, or cut and encroached into the common area of forest over the size of 3 or 4 *thước* [33m²] resulting in drought and damage to the rice fields?
>
> A: There are two or three houses that have rice fields near the forest, and they have cut some of the forest, but only one *thước*.
>
> Q: When there are papers and documents to be directed at villagers for their understanding, such as when you need to communicate to them about the laws of the government, who do you communicate with? Do you have meetings or post regulations?
>
> A: I just started working as village head in May last year, and I have not seen any paperwork, or when someone above me made paperwork or postings about permissions the village head has to carry out, they never allocated any paperwork to me, so then I have no idea about them.
>
> Q: If the villagers in your village do not know about the regulations, then where are the villagers getting their fuelwood or bamboo, wood, thatching, etc., to make houses?
>
> A: The area of forest around our village has two sides, one side is forbidden, one side is not, so we get our fuelwood from the forest that is not prohibited. People feel free to go there to get dry trees and bring them back for cooking, like bamboo, wood, thatching, rattans, but they have to

> ask permission of Mr. Forest Ranger at Yen Lap before they dare to go in, otherwise they will lose money.[37]

The feigned ignorance ("I just started this job," "No one told me about the laws," "No one sent me any paperwork") and sly resolution of problems ("One side of the forest has guards and prohibitions, so we go to the other side to avoid them") by this local official mirror the classic calculations of resistance and dissimulation long noted by James Scott (Scott 1985; 1990). This response also highlights how forest regulations themselves—which French authorities saw as the ultimate proof of their authority—could in fact become physical objects of resistance. One could protest that new regulations had not been passed down by authorities, or that posters had not been affixed in a public place, or that they could not be read if they were not written in vernacular languages or if locals were illiterate.

The increasing number of forest restrictions and new taxes and fees implemented after the 1914 and 1930 Forest Law revisions spurred new numbers of complaints. Forest tax revenues skyrocketed between 1910 and 1931, with the amount collected in 1931 nearly fifty times the amount in 1910 (Phan Xuân Đợt 1984). Many of these fees appeared to be arbitrarily set and enforced, and they often posed a significant financial hardship, particularly for the poorest. The reliance on taxes and permits for raising of revenue can primarily be explained by the fact that colonial administration in Indochina was largely enacted on the cheap, as "colonization must not cost France a thing" (Brocheux and Hémery 2011, 71). A letter from two-dozen Vietnamese resin collectors in Quảng Nam in 1934 to the president of the House of Representatives for Annam addressed this question of fees and their perceived injustices:

> We the undersigned, inhabitants of villages . . . of Quảng Nam have the honor to respectfully request you to kindly consider the following case: As mountainous region dwellers, who do not have sufficient land to work in our livelihoods, we enter the forest in search of resin (*dầu rái*). It is difficult to describe the dangers that we face in embracing this profession. The regions where we get resin are inhabited by *moi*s [savages], and before entering, we must pay them a yearly fee to avoid reprisals from them. Exposing our lives, on the one hand, to the malice of *moi*s, and the other hand, to the wild beasts, we run many risks in fetching the resin; leaving early in the morning from our house, we spend the whole day traveling to the mountains, and only after seven consecutive days of work can we

> collect resin, which nets us $1.50, or $1.00, even $0.80 or $0.70, depending on whether it was collected in good weather or bad weather. . . . With previous heads of posts [forest guards], we had to ask from them a simple authorization for penetration or movement in forests, but with the current head of the post, this is no longer so. Upon arrival, he made us understand that "according to the regulations, any person wishing to enter the mountains to look for forest products shall obtain a license as lumberjacks against payment of only $5.00." Alas, we are working as coolies for employers. Where do we find $5.00 for the payment of a fee for a permit? The wages we earn are barely enough to maintain our daily lives, our wives and children. Since the restrictions above reported, we have during the past two months been staying at home, doing nothing, creating hunger and misery. We would be very grateful if you would kindly intervene on our behalf with the Government that we are not obliged to be equipped with "permits" exclusively, and we can, through a simple authorization, go to work in the forests, as has happened in the time of the previous heads of the post.

The response from the Agent who was the subject of this complaint once again referred to the posting of laws as a sign of authority, as he simply stated to his higher-ups: "I have the honor to report to you that in the execution of my work as a forester, I am following exactly the instructions that are given in the October 1930 Forest Law, and so that no one can ignore this law, a notice in the Annamite language is displayed by me at the Ha Tan forest post." The petition for redress was denied.[38]

Given these situations, it was not surprising that more overt resistance escalated and occasional fits of violence broke out. Local peoples directed much of their ire against forest guards, considered at the very least inflexible, and, at the most, highly corrupt (Scott 1976). The inability of forestry officials to see the hardships from new forest restrictions, coupled with losses of other lands to agricultural concessions, added to the fact that the Forest Service had no money to support alternative economic opportunities that might replace lost forest income, created a powder-keg situation.

Social protests exploded in the spring of 1930, centered on the Nghệ Tĩnh area of northern Annam, and soon spread to nearly all parts of these provinces and even further afield, serving as a seminal turning point in the founding and subsequent military and political success of the Indochinese Communist Party (ICP). This protest movement has usually either been attributed to food insecurity caused by droughts and famines in north

Annam, which were further exacerbated by colonial tax extraction and a worldwide depression (a view identified with James Scott, for example), or to the important organizational role of the incipient ICP in the rebellious provinces (the viewpoint of Vietnamese historians [Nguyễn Trọng Cồn 1980]). Yet a forgotten historical footnote to these protests is the important role of forest resistance in contributing to the overall complaints.

While the protests originated among proletariat workers at timber factories, such as the Indochinese Consortium for Forestry and Matches (*Société Indochinoise des Forêts et d'Allumettes*) in the city of Vinh, a major forest exporting post on the Cả River, they soon spread to the countryside, culminating on May 1, a date chosen to coincide with International Labor Day (Đinh Trần Dương 2000). Workers in Vinh demanded pay raises, an eight-hour workday, and social security, while nearby peasants demand an abolition of the head tax, market taxes, forest taxes, river transport taxes, and a reduction of land taxes (H.T. 1980, 67). In one of the protests, between 500 and 3,000 peasants marched to the estate of a Vietnamese "collaborator" and sacked it, burning buildings and stealing grain, and shouting "accusations that he had stolen draft animals and prevented access to the forest" (White 1981, 60). By the fall of 1930, numerous villages in the province essentially governed themselves in so-called "Soviets" by establishing new rules on land access and tax collection.

For local authorities, there was but one explanation for all the unrest: communist infiltration that had lured weak and easily swayed peasants to their side. Whether or not these protests were truly spontaneous, or in fact directed by members of the Revolutionary Youth League founded by the young Ho Chi Minh, and the Indochinese Communist Party (ICP) that evolved from it, is a matter of considerable debate to this day (Trần Huy Liệu 1961). Certainly by the late summer and fall of 1930, the ICP Central Committee tried to take control of the spreading unrest and use it for recruiting. The Party issued directives to localities to implement various steps, including creating village self-defense forces, annulling all duties and taxes, confiscating communal fields from village "tyrants" and redistributing them, and forcing wealthy landlords to distribute stocks of paddy (H.T. 1980). In response, French officials in the region declared that by August the revenues collected up to that point were "enough" for the year, and no more tax collection would take place in the most restless villages. As the movement weakened, however, the French eventually reasserted control by brutal force.

Once calm had been restored, the problem of the "Red Terror" was extensively analyzed by the Security Police (*Sûreté*) (GGI 1933), and French authorities established a Commission of Inquiry into the Events in North Annam (*Commission d'enquête sur les évènements du Nord-Annam*) the following year to provide an in-depth assessment of the causes of the protests. One of their main findings was that the Forest Service had indeed been a contributing factor:

> The Annamese complain of too severe measures being taken against them for cutting timber and forests. Previously, the poor would cut wood and charcoal, which they sold. They thus had a source of income, which was decreased by the forest department. Currently those who cut down trees, if they are caught, are arrested and convicted. Similarly, they can no longer take advantage of wood to build their houses or do carpentry. That is the complaint.[39]

However, the commission members also argued that state forest restrictions had been reasonable, given excessive and unsustainable forest use in the area. The commission's final report read, in part:

> The Annamese is a great destroyer of forests. The hills and mountains were covered once with forests, but they currently have the appearance of leprosy. In all accessible locations, trees were cut or burned, and where there are still some forests, it was the difficulties of access that protected them. It is not uncommon to see at night a wave of fire moving up to conquer the mountain, destroying any hope of recovery of forest vegetation. To remedy this deplorable situation, the Forest Service began to reserve forests to spare them. Eight forest reserves, covering a total area of 122,000 hectares, have been created [in Hà Tĩnh]. Methodical cuts, granted to a contractor, have been put into operation. Nurseries were established, and the massif of Trân Trèo is in the process of reforestation. But the effort is insufficient. The forest service revenues are absorbed by the local budget. Part of the proceeds should be allocated to intensive reforestation. The forest is the rice of the poor, and gives fruit and food substitutes in case of famine. Scientific and rational exploitation would be a source of wealth for this unfortunate country. A miserable people would be in power and would find a livelihood and food.[40]

Yet a local mandarin and colonial official, Tôn Thất Trâm, who also sat as a member of the investigative committee, had a different interpretation of the events. Refusing to criticize local forest practices as being in need of control, he instead argued that colonial tax and land policy drove poor peasants towards higher forest dependence, setting into motion a vicious circle:

> Taxes are too high—there are many types of indirect taxes: taxes on forest products, seafood, marches, ferries, benefits, other funds, etc., that cause ruin to poor men of the people, such as the unfortunate who seeks his daily life in cutting firewood; or another who needs a few pieces of wood to build his house is forbidden and threatened by guards, and if he asks permission, he is taxed so exorbitantly that he can never pay.[41]

Ultimately, neither the recommendation of better organization and reforestation by the Forest Service nor reduced taxes for local people could be put into place, as the worldwide depression of the 1930s and eventual fall of French Indochina to the Japanese in 1940 ended all hopes for long-term state forest management under colonialism.

CONCLUSION

The French imperial project instigated major changes in how forests were seen, used, and managed. Viewing French colonial forestry as the early emergence of environmental rule helps us see how much of the work of problematizing and creating knowledge around forests resulted in specific forms of authority and interventions: the "how" and the "why" of rule. The identification of native forest practices as prime culprits for forest loss—rather than state mismanagement, or over-logging, or existing ecological conditions—served to create new relations and conduct in the countryside. Despite the fact that the earliest sites of French forest management, namely in Cochinchina around Saigon, quickly became some of the most degraded forests in the country, blame for deforestation was instead laid on mountain dwellers on the edges of the empire, with the identification of certain local practices, such as swidden or fire, as not only forbidden but a sign of laziness and errant subjectivity. Colonial paternalism towards these "barbaric" practices resulted in interventions and restrictions on forest use and attempts to change ethnic minorities' "culture," but in reality, there were no significant improvements to either. This stigmatization of the forest use of indigenous peoples remains a lasting legacy of the French era, and continues to lead to tensions between upstream and downstream areas to this day, particularly since most subsequent forest policies have insisted that upstream communities must act differently for the protection of water users downstream, despite little evidence to suggest that such a focus would be effective in securing regular water supplies.

The calculative mechanisms used by French foresters to justify their expertise and authority included surveys and inventories to assess the extent and type of available timber and the identification of species to be regulated in trade and use, and these new forms of knowledge translated directly into the formation of a state Forest Service, manned by "expert" foresters trained in the metropole. Despite the clear prioritization of timber production as the primary goal of forest administration, officials often invoked "scientific" knowledge, particularly regarding water control and climate, as a prominent justification for colonial forest control. Yet the interventions in forest areas, such as the creation of so-called "Reserved Forests," were not truly about environmental problems like flooding or soil erosion, as much as these reasons were presented. Rather, it seems clear that the financial considerations for forestry dominated, with management aimed at producing a minimal supply of timber and taxes for a sufficient departmental budget. The location of the Reserved Forests provides the clearest evidence that the Forest Service remained primarily motivated by profit, not by environmental concern; the visualization and mapping of reserves were not based on ecological criteria or rarity of species, but on what areas were potentially accessible and economically valuable.

Networks served key roles in spreading knowledge and authority about environmental rule as well. From an initial circle of a very small number of French-trained European foresters, the colonial Forest Service emerged, and French foresters tried to persuade other actors, such as hydraulic engineers, traders, and Vietnamese mandarins, to adopt their viewpoint that forests required stricter controls by invoking concerns about flood control, export material, and land cover, respectively. Yet these forestry networks did not only run one-way from the colonial powers to the colonies; knowledge became localized and specific to Indochina, and circulated back out, like the collective punishments for fires in the Indochinese Forest Law that were extended to other French possessions. Objects themselves also circulated through these networks, like the 1669 Forest Ordinance that moved from the countryside of the motherland out to Indochina, where it was revised and applied, and from there outward to other colonies in Africa. But sometimes the circulations stopped, or were resisted, like the handbills that spelled out the new forest restrictions for local communities, which, once posted in community areas, were often ignored or protested against.

Yet despite the numerous actions that comprised the network and regime of practices of colonial forestry, French management was not nearly as complex and regulated as that of the British in India. Oftentimes, management appeared desultory, and continued budget crises eventually lead to a laissez-faire treatment of the forest sector as a means to raise revenues through taxation. Ecological conditions exacerbated the situation by simply refusing to produce expansive stands of valuable timber, making organized exploitation prohibitively expensive. This "resistance" by trees themselves made it much easier to base management on the regulation and taxation of private concessionaires and loggers than to build a significant forest service from the ground up. Management that was supposed to be progressive and rationalized was thus restricted to less than 2 percent of the forest estate in most areas, and even there, successes were elusive.

In spite of the many failures, there is no doubt that the French colonial project succeeded in introducing new ways of thinking about forests into Indochina. Similar to "political forests" elsewhere in Southeast Asia, colonialism first introduced the idea of the state as the sole managerial authority for the best and most valuable forests, an idea to be continued in later years of environmental rule under post-colonial regimes. Yet the incomplete authority by which forests were assigned to the state, combined with weak personnel forces, provided little incentive for long-term management either for local communities or commercial enterprises. Not surprisingly, this state project clashed with peasant demands for continued local control, based on historical and moral arguments, and the protests of the Nghệ Tĩnh Soviets showed that resistance to top-down state environmental rule could be successful, a lesson which was to be repeated through Vietnam's forest history.

2 Planting New People

Socialism, Settlement, and Subjectivity in the Postcolonial Forest

WITH THE DEFEAT OF THE FRENCH IN THE POST–WORLD WAR II period, a new era for forest management arose in colonialism's place. The rupture from the colonial past happened on a date firmly enshrined in Vietnamese history: September 2, 1945, when Ho Chi Minh, leader of a small group of anti-colonialists called the Việt Minh, declared independence. As noted previously, Ho Chi Minh himself recognized the strong role that the environment and forests could play in the development of a modern nation. The emergent postcolony of the Democratic Republic of Vietnam (the DRV, or North Vietnam), struggling against their enemies in the South, and later against US involvement, needed to build a new state that met their collectivist ideals. Land reform and redistribution and collectivization of agriculture were key policies to boost food production. But these were steps largely undertaken in delta areas, by farmers already engaged in intensive agriculture. More needed to be done to bring large-scale socialist development to the highlands and forest areas that made up 70 percent of North Vietnam's land area.

True to environmental rule everywhere, problems were "discovered" in the new DRV, and policies made to fit them. Forests became a canvas for the design of large-scale political projects. Social problems—such as how to integrate ethnic minorities into a majority state, and how to turn peasants into socialist workers—became "environmental" problems. Forestry was problematized as a field that needed to quickly produce massive quantities of wood and thus needed to be organized and run by a central state. Ethnic minorities who already resided in the mountains were problematized as untrustworthy people producing food inefficiently through swidden, who needed to be guided by the party to more rapid and scientific progress, preferably by settling in state-organized villages. Thus, the "natures" and landscapes that were prioritized in policy reflected concerns about people more than forests and ecology, as the construction of the new DRV required the production of new citizens at an important time of state formation, particularly in frontier areas (Lentz 2011). One important but overlooked part of this process was the use of forest environments, and peopling these areas through migration and resettlement, to create new subjectivities aligned with the goals of the state. As Ho Chi Minh himself noted, "If you want to have benefits for ten years, plant a tree. If you want to have one hundred years of benefits, then plant people."[1] Like the process of reforestation, people too were "planted," moved to new physical locales as objects needing state guidance, and nurtured to become new citizen-subjects in particular ways.

Environmental rule was enacted primarily through two specific technologies to shape the "conduct of conduct" in the DRV: the creation of state-owned logging companies (called State Forest Enterprises [SFEs]) helmed by new socialist citizens, working and living together as a novel type of family, and the resettlement of swiddeners into sedentarized, state-planned villages. Massive relocation and migration programs moved nearly a million lowlanders into the mountains from 1960 to 1975 to work in new highland ventures, while at the same time attempting to resettle nearly two million mountainous minorities into state-planned locales. New SFEs were manned by these lowland recruits, the majority of them women, who arrived in the highlands and began new lives as forestry workers in what planners hoped would be an advanced socialist paradise in the hills, while sedentarization projects moved ethnic minorities into concentrated settlement sites where they could be "improved."[2] State-led environmental rule during this era thus had as its primary underlying goal the creation of new citizens from forest work, rather than the protection of forests from people.

In addition to the social revolutions that emerged from the DRV era, ecological changes further transformed the highlands. Under the DRV, forests were reimagined as zones where large-scale scientific socialism could be put into practice. Assistance from East German and Soviet foresters reinforced the idea that Vietnam's forests could be like those countries' intensively managed temperate woods producing timber with regularity for the state. Yet neither Vietnam's environment nor its people quite fit this industrial socialist dream. The DRV's forests were heterogeneous and could not produce timber sustainably in large enough quantities, and its people, heterogeneous as well in ethnic diversity, were engaged in daily struggles for subsistence, not perfection as socialist models. Rapid transformation and the degradation of many highland environments were an inevitable outcome of the challenges of wartime damage, overestimation of the amount of timber forests could supply, the poor planning and over-bureaucratization of socialist enterprises, and the lack of attention to more sensitive ecosystems for conservation purposes.

Data is not available to make conclusive statements about rates of forest change with precision during this era, as the existence of "forests" depended on how officials and ministries described and understood different landscapes. Some areas with trees were called "forests" and were expected to produce timber, while other areas with trees were called "wastelands" and were expected to be converted into agricultural fields. Many times these categories were fluid and highly subjective. We can, however, focus on how "nature" was represented by those in charge of these various programs of environmental rule. New definitions and classification of forests led to particular technologies of intervention, namely the SFEs and resettlement sites. This new bounded entity called a "forest" was enacted through the articulation of knowledge production and practices such as surveys, logging quotas, and labor practices; objects such as chainsaws and Molotova trucks; resettlement sites as new modern communities; and environmental responses from trees themselves. In turn, the new reimagined forest spaces helped make new reimagined citizens.

WAR, FAMINE, AND THE BIRTH OF A NATION

In the summer of 1940, France had fallen to Germany, and the new Vichy regime was installed, with nominal control over French overseas holdings. Soon thereafter, the Imperial Japanese Army moved towards French Indochina, and an agreement with the Vichy regime allowed for a limited Japanese occupation in return for avoidance of hostilities (Jennings 2004). Indochina served as a source of raw materials for the Pacific front during World War II, with serious effects on the local environment; an American report complained that 50,000 hectares of what had once been French reserved forests "have been devastated by the Japanese, who cut down all the fine trees without consideration for the species of trees and the future of the forest" (Brown 1957, 45). Against this backdrop of political turmoil and war was a further tragedy. A ferocious famine struck across the North of Vietnam in 1944–45, killing an estimated one to two million people, nearly 10 percent of the population (Marr 1995), and the memory of the famine would later strongly motivate concerns with land reform and agricultural expansion. The global peace following the Japanese surrender in 1945 did not prevail in Indochina, which soon descended into the eight-year Franco-Vietnam war.

The defeat of French forces at the battle of Điện Biên Phủ in 1954 brought a temporary cessation of hostilities to North Vietnam and recognition of an independent DRV, although the state had begun to take shape before this moment. The most well-publicized transformations of the rural countryside were extensive land reform policies passed in December of 1953, directed at "seizing land from French colonizers and other invaders, and erasing the feudalism of the landlord system."[3] Overall, approximately 700,000 hectares of land were redistributed to poor and landless peasants throughout North Vietnam at this time (Moise 1983). The cooperativization of agriculture was the next potential step to achieve the goal of national socialist development aimed at food security (see Kerkvliet [2005] for an in-depth exploration of this process). These cooperatives succeeded in areas where rice farmers had long required collective action to regulate water supply and access. But what about highland areas? Most of these were forests, scrub, grasslands, or other so-called wastelands. How were they to be brought into socialist development? Party tracts emphasized the need for mountainous areas to establish cooperatives as well, but the challenges were great when populations were not physically concentrated as in deltas, and with spatially extensive agriculture, as was the case with swidden. By 1962, while it was estimated that 80 percent of delta and valley farmers in the DRV were in some form of cooperatives, only 20 percent of highland farmers had been motivated to join (Chu Văn Tấn 1962).

Thus it was clear that in many environmental zones of the mountains, cooperatives would be insufficient; rather, state-led development (*kinh doanh quốc gia*) would be key and would be achieved through state agricultural farms (*nông trường*) and state forest enterprises (*lâm trường*). Most importantly, these new state farms and forest enterprises required the movements of large numbers of people from coastal and lowland areas to the "underdeveloped" hills and along sensitive border regions, as clearing land for these new farms would require significant labor and mechanical power (Hardy 2005). The first Five Year Plan of the DRV (1960–65) ambitiously made targets for the clearing of 200,000 hectares of land by state farms, and an additional 350,000 hectares by cooperatives (VWP 1963).

KNOWING, MAPPING, AND GOVERNING THE FORESTS OF THE NEW NATION

As was the case for agriculture, DRV leaders promoted a completely new approach to forestry; new institutions needed to be set up to regulate novel practices. The Việt Minh founded a Ministry of Agriculture (*Bộ Canh nông*) in 1945, with a smaller division of forestry (*Sở Lâm chính*) under it. Forestry was promoted to an equal place when it was renamed the Ministry of Agriculture and Forestry (*Bộ Nông lâm*) in 1955. Under the national ministry, each province created a provincial department of agriculture and forestry as well (*Ty Nông lâm*). Within the national ministry, departments such as the subdivision of National Forest Clearing Enterprises (*Chi nhánh Quốc doanh Lâm khẩn*) provided an indication of the importance land clearing, migration, and reforestation would come to play in the new forest strategies of the DRV.

The first new forest regulations for the DRV were issued in 1956, just two years after the official founding of the nation.[4] But, in reality, authorities did not have a clear handle on what the resources of the new territory were, and what could profitably be exploited. The loss of the rich forest estate of the south, now under the control of the Republic of Vietnam (South Vietnam), weighed heavily, and so officials eagerly sought out assurances that the North too had sufficient supplies. Institutions for knowledge production were formed, including a new State Committee on Science (*Uỷ ban Khoa học*), charged with sending out survey parties to various provinces and reporting back on physical and natural resources. A new national forest survey office (*Chi nhánh Điều tra Quy hoạch rừng*) quickly hired thirty people to undertake a detailed inventory, and from 1956 to 1959, 1,140,330 hectares of forest were ground explored and mapped.[5] Their survey, completed in 1959, revealed sobering news: the only areas of "rich" forest left in the country were around the Ba Bể area of the northern mountains and in the provinces of Hòa Bình and Nghệ An in the former North Annam (see map 2.01). Large areas of "poor" (*nghèo*) forests and "bare hills" (*đồi trọc*) were identified throughout most of the northeast, and along the coast as far south as the seventeenth parallel.[6] Given this news, the Forest Service decided the easiest classification system for the new DRV's forests would be based not on ecology or climate, as other systems in use at the time were, nor on the older French system of forest classification according to timber quality. Rather, the Vietnamese were influenced by the writing of an East German forester, M. Loschau, who had noted that presence of human influence could be the basis for forest classification. Based on this, four main kinds of forests were identified for the DRV: one, bare hills in need of reforestation; two, young forests in need of supplementary planting; three, secondary forests with some exploitable timber; and four, mostly primary forest with substantial timber trees. Authorities advocated the use of this system because it was very easy to use in the field through visual assessment by non-specialists (Thái Văn Trừng 1970). A visit to the DRV by Loschau to see his classification system in action revealed around 50 percent of the forest area was category one (bare hills

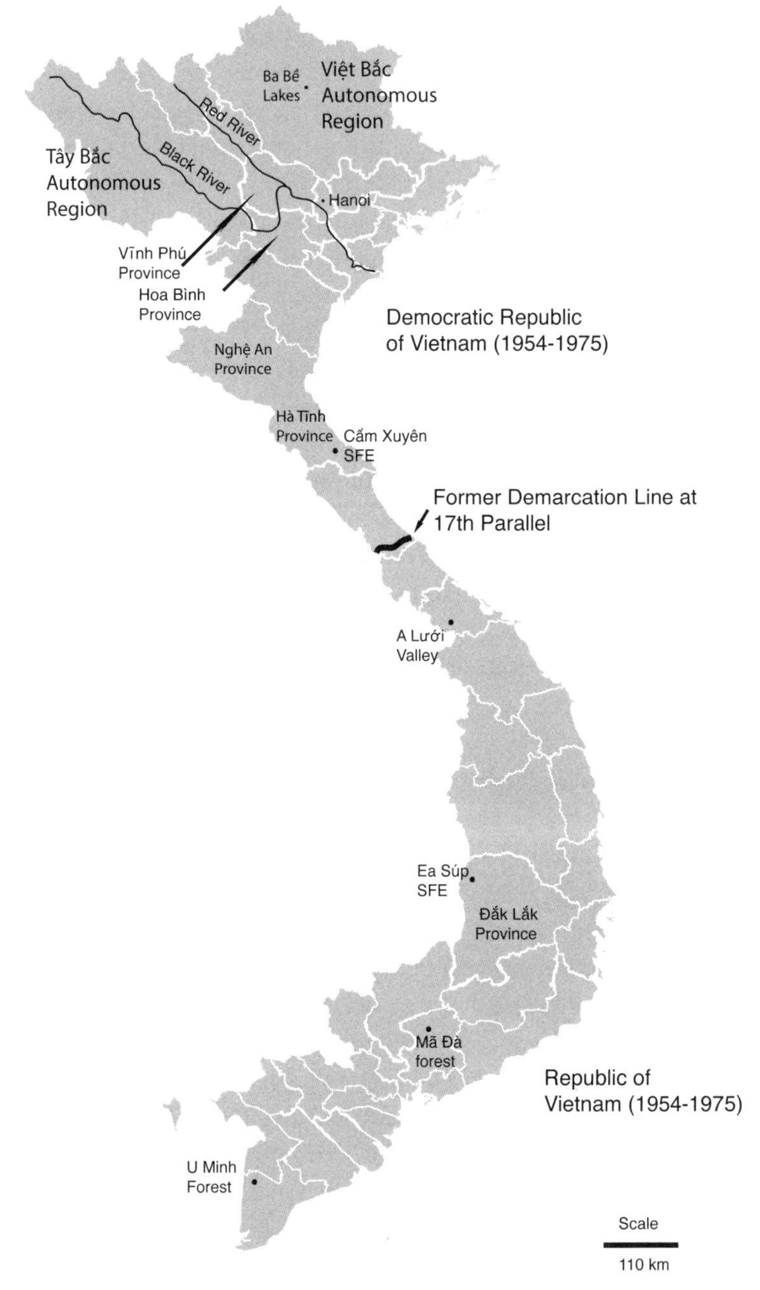

MAP 2.01 The Democratic Republic of Vietnam (North Vietnam) and Republic of Vietnam (South Vietnam), 1954–75, indicating places mentioned in text. Base map from Vietnam location map by Uwe Dedering on Wikimedia Commons. Map by author.

and deforested clearings); 25 percent was category two and three, and "only about 25% of the forest area consists of little or unused forests with sizable usable timber supplies. These forests are, however, mostly in mountainous areas inaccessible for forest management" (Loschau 1963, 118). With forest problems now identified (namely, the lack of exploitable trees on half of the "forest" land area, and best quality timber remaining in inaccessible areas), and new institutions in place, the large-scale development of the rural and mountainous areas of Vietnam could begin.

STATE SOCIALISM: REGULATING THE FOREST

In building a new forest policy, DRV officials were sure of the direction they did not want to take: back to the colonial past. French forest management was critiqued as haphazard, inconsistent, and exploitative in order to enrich the motherland, rather than to develop a robust national economic sector. An early report in the Vietnamese archives, likely from 1955, explains the thinking of officials:

> In the time of French colonialism, because exploitation had no direction (the policy of colonialism was just short-term exploitation) the work of management was very light. The French Forest Service was only a tax service, and did not perform any specialized duties to manage forest products, or protect forests. Because of this, a lot of Vietnam's forests were destroyed; places that used to be rich in materials and easy to exploit have been exhausted. Lots of valuable forest products with many uses were not protected, were not developed, and were exploited recklessly and gradually have become more rare.[7]

The fact that the forests had not been used to enhance the development of the nation as a whole was a particular sore point, exemplified by the French practice of auctioning off rights to harvest reserved forests to private companies and concessionaires, as well as turning over forest areas to agricultural and rubber companies. DRV officials further worried that the mindset of people living near forests had hardened during the colonial period, and it was going to be difficult to get them to recognize the need for a new management approach:

> [T]he bad impacts of the lack of responsibility of the French creates many difficulties. For people, the perspective the "forest is endless" [*vô tận*] is

> quite serious: individuals exploit forest products, local offices exploit forests directly without recognizing the need for management, and are not yet meeting the requirements for protection of forests . . . we need to fight against this idea that forests are endless, and educate cadres and citizens to pay attention to the point of view of protection and development of the benefits of forests.[8]

The difficulties were particularly pronounced when it came to recognizing the authority of new forest officials and guards, as under the French these agents were considered corrupt, and the restrictions they imposed on local use viewed as arbitrary and unreasonable. This left behind "a difficult impression to fade away in the consciousness of our people about forest rangers, that rangers are corrupt and oppressive (*tham nhũng và ức hiếp*) to the people: [they say] 'first [money] to the rangers, second to the governor of Indochina' (*nhất kiểm lâm, nhì khâm sứ*)" (Phan Xuân Đợt 1984, 42). The task of the new forest administration was to get forest-using peasants to think first about the national interests of the DRV rather than their own needs.

The initial prognosis was not promising. As early as 1946, the Việt Minh pleaded for attention to forest management as a responsibility of the state.[9] Scattered efforts to organize forest activities throughout the war years were attempted, but timber production dropped precipitously from the heights of French and Japanese exploitation in 1942, with only around 1,000 m^3 produced in 1951 (see table 2.01). One immediate concern, given the low production of wood during the turmoil of the early resistance, was that even humble technologies like paper and matches were in short supply. While the Việt Minh constructed their own rudimentary paper mill in Vĩnh Phú province in the early 1950s, shortages continued throughout the early days of the new regime and had an impact on the kinds of work that could be carried out. The paper shortages were seen as not only a failure of economic production but potentially of educational production as well, as having access to paper helped generate "culture and civilization" through the propaganda work essential to the new state (Pelley 2002).

When a series of new DRV forest regulations came out in the mid- to late-1950s, there could be no mistaking the direction of the policy: "Forests are a national asset that the State manages" (*Rừng là tài sản quốc gia do nhà nước quản lý*).[10] But the "State" was not yet present in many rural areas, and so local areas with forests were each to set up a "board for forest regulations" (*ban quy định rừng*), comprising representatives of communes, members of farmers' associations, and representatives of people's committees. This forest

TABLE 2.01 Total timber production by year, 1939–54

Year	Total m3 Produced
1939	277,673
1942	364,220
1948	92,501
1949	40,986
1950	21,453
1951	1,008
1952	23,710
1953	20,431
1954	81,000

Source: NAV3, Bộ Nông Lâm files, folder 5622: Báo cáo thông kê tình hình sản xuất lâm sản 1958–59 [Report on statistics of the situation of forest product production].

committee was to organize local people to assess the state of nearby forests and to lay out the borders of areas needing regulation.

This loosely organized, mostly decentralized, forest policy lasted but a few years and was essentially abandoned before it could be implemented. By 1959, authorities were already concerned that this system was not leading to the development of "socialist forestry, aimed at meeting in time the needs in front of us and in the future"; timber cutting was "not yet organized appropriately and there is still much wastefulness." Moreover, "forest fires are still widespread, swidden has not yet been guided closely, and the work of reforesting is still limited."[11] Instead, authorities moved more closely toward a new centralized approach that had already been tested in agriculture: the development of state-run farms and forestry enterprises. The first Five Year Plan aimed to supply raw materials:

> If we want to develop and industrialize in the socialist way, we cannot consider mountain forests occupying half the area of our north as just a natural storage of forest goods that is invaluable and of immeasurable diversity for a tropical country like ours. If we want to raise the standard of living for our people, we also cannot simply consider it sufficient to have exploitation activities and processing of forest goods limited to just three

> million minorities of ours in the mountain forests, where forest goods collection can occupy more than 60 percent of the harvest each year even when these people are in agricultural cooperatives.[12]

This turn to remake the forest sector for industrial development moved Vietnam towards the path of Soviet and East German models. Discussion at the time focused on an ideal of four phases of industrial forestry: first, open exploitation; then, protection; then, replanting in even aged and single species stands; and finally, the complete industrialization of the forest estate manned by trained specialist workers according to scientific and Taylorized methods (Nguyễn Tạo 1968). Yet DRV officials noted the tropical climate made Vietnam "different than other socialist countries," as in Eastern Europe, and an industrial plantation economy would need to have clear designations between forests, agriculture, livestock, and industry, as the latter alone would not be sufficient (Anon. 1968). Yet this approach failed to recognize Vietnam's ecological situation as one with high forest species diversity yet low timber quantity, a problem plaguing all previous efforts at French forest management, and which would not lend itself to industrialization either, as the types of large-scale wood processing enterprises successful in the Soviet Union and East Germany relied on consistent quantities of one or two types of wood. With Vietnamese forests' low amounts of timber and high levels of variety, it would be a major challenge to raise enough revenue to industrialize this sector. Despite these considerable problems, however, the state moved to further consolidate the nationalization of the forest estate and the establishment of SFEs (*lâm trường*) to log and manage these lands as the first step toward a glorious socialist future.

INTERVENTIONS: CREATING STATE FOREST ENTERPRISES

Lâm trường, like many Vietnamese terms, is porous semantically. *Lâm* is a root word for forest, while *trường* can be a school, a field, or a public works project.[13] The SFEs encompassed all three of these aspects. The first formally organized "forest enterprise" called a *lâm trường* had been set up in 1950 in the Việt Bắc war zone to supply charcoal to the war effort. As SFEs were extended into new areas, priority was given to areas with sufficient transportation networks to move logs, sites relatively close to consumer centers, and areas relatively populated so that sufficient labor was available.[14] In this, the new DRV's forest priority areas looked a lot like the old French reserved forests, which were established based on accessibility and ease of logging, rather than ecological or other criteria. This process of site selection served to replicate the unevenness of state forests begun under the French, whereby certain forests, such as mangroves or scattered open forests, received little to no attention, while upland forests, especially forests dominated by key species like pines, were heavily targeted for logging.

From these beginnings, the SFEs began to take shape and continued to spread across North Vietnam, formed as production units under district or provincial levels of administration, although some especially valuable or strategic SFEs were directly operated by the central government (known as *Lâm trường Quốc doanh*).[15] Most SFEs were structured along much the same principles as the French had envisioned for the reserved forests; they would be surveyed and an annual allowable cut developed based on inventory. Stands would be logged regularly, under clear-felled rotations of twenty to thirty years, to create evenly aged forests. Every year targets would be set for wood cuts, based on both local inventory and national projections of needs, and, as parastatal companies, each SFE also had a target financial "contribution" to local or national budgets they were to hit each year. Particularly in remote, sparsely populated areas, forestry organized by SFEs was envisioned as a driver of economic activity and source of employment. In these areas, the SFEs were not only logging companies, but often the only sign of the state itself: they constructed roads and supported village development by installing water supplies and even operating schools.

We can follow the development of SFEs through the transformations in a single province, Hà Tĩnh. There, six state enterprises were set up before 1960, including three state agricultural farms and three SFEs (Cẩm Kỳ, Hương Sơn, and Đức Thọ; see map 2.02). Cẩm Kỳ SFE had an area of 10,290 hectares and was established on land previously designated a French reserved forest to supply railroad ties. (Cẩm Kỳ is of significance because it would later became known as the Cẩm Xuyên SFE in 1972, and it would eventually be converted to the Kẻ Gỗ Nature Reserve in 1996.) One of the SFEs in Hà Tĩnh was directly under the national government, known as Hương Sơn SFE (*Lâm trường Quốc doanh Hương Sơn*), comprising 78,000 hectares in the foothills of the Annamite Mountain chain. It too was located on a previous French reserved forest, taking advantage of the Ngàn Phố river so timber could be floated down to reach the port city of Vinh/Bến Thủy.

However, the Hương Sơn SFE was in a very remote area with few local laborers. Thus, the organization set about recruiting 1,500 new workers from the coastal areas of the province; given a shortage of land and high poverty in coastal districts, many young recruits were initially eager about

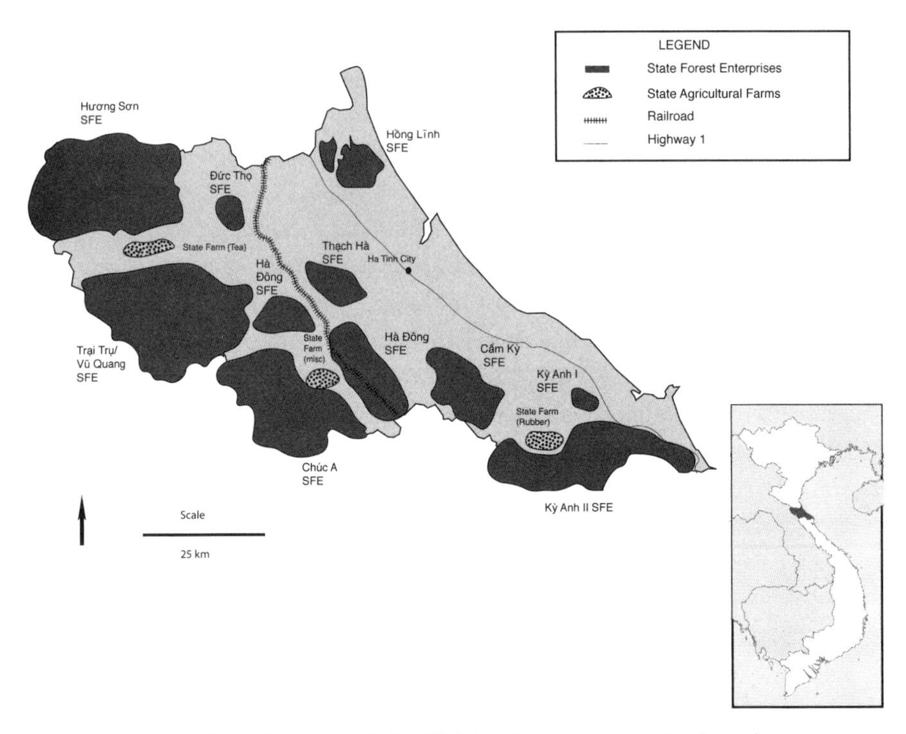

MAP 2.02 State Forest Enterprises in Hà Tĩnh Province, circa 1975. Map by author.

jobs, but many were "afraid, worried, and unsettled about this area," fearing it contained "poisoned waters and malevolent spirits" (Văn Ngọc Thành 2005, 29). A number of the recruits fled in 1956–57 due to poor organization and a lack of food, and by the end of 1957 there were only three hundred workers left. In the first full year of operations at the SFE in 1958, forestry equipment was provided by the Soviet Union, including Molotova trucks and steam-powered chainsaws. By the early 1960s, Hương Sơn had four hundred workers, divided in fifteen brigades for lumbering, four road crew brigades, four tree planting brigades, two transport units, two processing units, and one mechanical repair unit. The yearly targets in the 1960s were to cut 30,000 to 40,000 m^3 of wood each year, and to replant 200 to 300 hectares (Văn Ngọc Thành 2005).

Eventually, by the mid 1970s, nearly all accessible land in the DRV with any semblance of tree cover was put under control of an SFE in a massive process of forest nationalization; in Hà Tĩnh province, for example, the original three SFEs expanded to ten by the late 1970s. These local enterprises, together with SFEs in the neighboring province of Nghệ An (with which Hà

Tĩnh was sometimes conjoined) supplied approximately 250,000 to 300,000 m^3 of timber and 150,000 to 200,000 m^3 of firewood each year by the mid-1960s (Nguyễn Phúc Khánh and Nguyễn Văn Trọng 1986). The massive scope of these operations can be compared with previous timber cuts; for example, in 1929, the French Forest Service reported only 25,000 m^3 had been cut from all "Reserved Forests" in Indochina, while in 1963 these two provinces alone were cutting ten times that amount. The huge expansion in forest exploitation across the DRV can be seen in statistics: using 1955 as baseline, in which 361,380 m^3 of wood was produced, by 1958, the DRV had increased wood production by 125 percent; by 1960 they had increased by 199 percent, and in 1964, they posted a 300 percent increase, with 1,108,693 m^3 total production for the year (Nguyễn Tạo 1968, 34).

Not surprisingly, with this rapid expansion, early troubles were already plaguing the new SFEs. Inventories of the local forests often only took place *after* the SFEs had been set up and begun operations, and SFEs were said to be "exploiting erratically and not according to established processes, not preparing forests, not cleaning forests, cutting timber recklessly and endangering forests, all of which makes the forests exhausted very quickly."[16] Despite the widespread damage, yearly targets for forest production were not being met in many areas, and wood loss from poor cutting and processing was estimated at 40 percent of the total amount of trees felled (Loschau 1963). There was minimal attention to actually assessing the quality and quantity of forests under SFE control before target plans were made each year. In fact, in reports that SFEs were to submit to the General Directorate of Forestry in Hanoi each quarter, there are few maps or sketches of forest areas and inventory, and almost no mention of specific species or felling plans. Instead, most of these documents focus on the minutia of points systems for awarding salaries, or assessments of equipment, even down to the mileage on SFE-owned vehicles.[17] (In a complaint letter from the head of the General Directorate of Forestry to a local SFE in Ninh Bình province, he accused SFE officials of having "better cataloguing of auto parts for cars in your SFE than good tree inventories."[18])

This lack of attention to tree growth rates or species composition was chalked up to a lack of expertise. For example, in 1960, the Hương Sơn SFE employed only fifteen people who had gone to secondary (high) school, let alone a tertiary forestry education, while the rest of the three hundred "technical" cadres had not attended any sort of secondary school (Văn Ngọc Thành 2005).[19] In order to address these problems, the Forest Science Institute of Vietnam engaged several foreign experts from friendly countries to come visit the DRV and assist them in professionalizing the SFEs and helping them make scientifically sound cutting and management plans. East Germany provided assistance with forest inventories; Soviet and East German scientists assisted with soil research, tree physiology, and wood technology (Loschau 1969); and entomologists from Hungary also visited (Westing 1974).

The East German consultants who arrived in the early 1960s to help carry out inventories were dismayed at what they found. SFEs had no sense of what trees were valuable and how they should be used. The German foresters noted that the highest quality woods, those labeled as "classified species" during the colonial era, were by the 1960s almost completely exhausted, leaving forests filled with species for which there was no economic interest. Because the species richness made standardization of timber difficult, the Germans recommended that focus instead should be put on uses of wood other than timber (Harzmann 1964). They particularly recommended that the best solution would be the "mechanization and rationalization of all forest works" through use of "widely different species of tropical forests, in particular for the production of fiber, paper, plywood and pit-props" (Loschau 1963, 119).

Whether the Germans' advice was the cause or not, forest production did continue to increase after the early 1960s. One of the largest reasons for the increase in production, however, was not just new technologies, but expanded labor. As SFEs spread across the mountainous landscapes of the north, more extensive recruitment for workers became a pressing concern, which the state solved through extensive upland migration plans. In making this move, both the peasants themselves and the forest environments on the edges of the new nation were transformed.

NEW SUBJECTIVITIES: RECRUITMENT, MIGRATION AND BECOMING A "*ĐỒNG CHÍ*" (COMRADE)

Relocating lowland recruits to join the SFEs as employees and move to upland and forested areas was just one of several overall resettlement plans formulated by the DRV to move almost a million ethnic Vietnamese (Kinh) from the deltas to uplands (Hardy, 2005). These resettlement programs had first been suggested under the French, but they had moved very few people. The DRV leaders were more strongly motivated, as officials worried overpopulation of the Red River Delta would lead to the food shortages they had so long feared after the 1945 famine. Thus they announced resettlement plans to reduce the population of the Delta by 70,000 people a year through out-migration (Lê Hồng Tâm 1972). The hope was that this redistribution would even out population densities, from over 1,200 persons per square kilometer in the Red River Delta to fewer than fifty in the uplands.

One such program to move peasants was titled "Clear the Wilderness" (*Khai hoang*), which sent lowland agriculturalists to the uplands to clear lands for cooperatives and state agricultural farms; another later program was titled "Building New Economic Zones" (*Xây dựng các vùng kinh tế mới*). These migrants received assistance from the state in the form of collective residential and agricultural land and tools of production, but house construction, clean water, schools, health centers, and other forms of infrastructure often had to be built by the settlers themselves through cooperative work groups. Although employment in SFEs was a much smaller project than "Clear the Wilderness" and "New Economic Zones," thousands of lowland peasants were recruited for these operations as well. Those who signed up for SFEs had a different job than the Clear the Wilderness and New Economic Zones migrants, one they considered more stable and desirable work, because they would be provided with state incomes, food, clothing, and housing at a minimum. Further, many SFE recruits hoped state employment would be temporary, and they could one day return to their lowland natal villages, while other migrants often sold land and everything they owned in their sending village when they left. The SFE jobs often did not provide an option to bring a family along, as the Clear the Wilderness program did, and so recruitment for SFEs often aimed at the young and unmarried. Although SFEs did not specifically target female labor, because young men were off at war in the 1960s and 1970s, they had disproportionate numbers of recruits who were women. In many SFEs, by the late 1970s, some 70 percent of workers were female (Trung tâm Nghiên cứu Phụ nữ 1987).

For SFEs, there were obvious advantages to recruiting in lowlands. Most workers recruited were not trained in forestry; the majority were simple manual laborers. Because rice cultivation in delta areas made use of skills SFE workers would need (such as the willingness to work long hours in the elements, the ability to haul water and soil, and experience in raising plants in seed nurseries), the agricultural peasantry seemed a useful labor pool (Liljeström et al. 1998). Further, bringing workers up to SFEs from the deltas was a way to procure reliable employees who would be fully committed to the SFE, as their families would be far away, and thus they could be counted on to provide consistent labor. This was contrasted with the local populations already living around SFEs, who were for the most part ethnic minorities; most managers of SFEs thought minorities would be unwilling and ineffective workers (Fortunel 2009). For many years, until the early 1980s, most SFEs refused to recruit or hire local ethnic minorities.

Becoming an SFE worker differed from working in a cooperative or being a Clear the Wilderness migrant in significant ways. SFE workers were state laborers and thus subject to rules and regulations on wage rates and compensation; cooperative members on the other hand would receive food from the cooperative's production, depending on the number of points they had accumulated or land they had donated. SFE workers were called "laborers" (*công nhân*)—the same appellation given to industrial factory workers—and were paid on state contracts (*hợp đồng*) with access to retirement benefits, housing, and food coupons. Labor in SFEs was assigned by a point system, and food and other necessities were then bought every month with coupons.[20] Work points were awarded by activity, area, and time spent in the forests. For example, table 2.02 shows the ranking of forest labor; those in the top categories were considered to be doing the hardest work and would collect the most work points, although nearly all work was done by hand given low rates of mechanization. Workers were expected to be on the job twenty-six days each month, although absenteeism was a serious problem in some areas (Nghiêm Thị Yến 1994).

Life in an SFE was all part of the process of making new socialist citizens. Across the DRV, citizens were urged to make themselves into the "new person" (*con người mới*) in a socialist "new society" (*xã hội mới*) through reformation of everything from work and family life to ritual and customs.[21] SFE work was just one among the many tools used to remake subjectivities and conduct. SFE workers took to the new practice of referring to one another not by age and gendered pronouns typical of spoken Vietnamese, like "brother," or "senior aunt," but rather as "comrade" (*đồng chí*). To further the sense of camaraderie, most SFE workers lived in cooperative row housing in cleared areas near the forests they were to be harvesting. SFEs often resembled small villages or towns, as they had to provide all of the services that workers might need, including schools, canteens, health facilities, and entertainment. For example, the Hương Sơn SFE in Hà Tĩnh had grown from just a handful of workers at its founding in 1955 to 4,000 by the mid-1960s. The SFE had a nursery, an elementary and middle school for the 1,300 children living there, a comprehensive hospital, and fifty collective kitchens where workers would eat each night.

Entertainment on an SFE was often difficult, but necessary for the workers to endure the hardships of being far away from family and friends and in a new environment. The SFEs encouraged collective exercise each morning

TABLE 2.01 Examples of labor activities in an SFE, ranked by difficulty

1	Collecting seeds for planting; climbing tall trees; skidding logs by hand; laborious work with rocks
2	Workers leading buffalo to skid out wood products; other laborious hand work
3	Workers in forest who work with buffalo, or elephant mahouts
4	Workers sawing by hand in the forest
5	Workers using electric saws
6	Workers who regularly have to actively enter the forest for manual labor [eg, for hoeing, weeding, etc.]
7	Workers who remove bark from logs
8	Workers who weed mechanically
9	Workers who shellac wood
10	Workers who work with gravel
11	Workers who catch wild animals
12	Workers who spray pesticides
13	Workers who apply fertilizers

Source: NAV 3, Tổng cục Lâm nghiệp files, folder 11 LN/TT: Thông tư bổ sung thông tư số 16 NL/TT ngày 27/4/1959 của Bộ Nông lâm.

and organized friendly competitions between different brigades based on days of fire protection, acres planted, or other metrics.[22] The Hương Sơn SFE, which was near the beginning of the supply route for the Ho Chi Minh Trail, was under constant threat of air attack, and the ongoing war made watching TV or playing outdoor soccer impossible, as electricity or group movements might attract a US aerial raid, so instead workers were encouraged to write articles for newspapers or read poems by lamplight. A theater troupe entertained occasionally with works from literature (Văn Ngọc Thành 2005). Yet even these attempts at entertainment could not hide the fact that, as a research report on SFEs later put it, "[A] worker's life is completely isolated from outside social information" (Nghiêm Thị Yến 1994, 6). It could take as much as a day to walk to a nearby town to buy a postage stamp to mail a letter. Although SFEs often provided vehicles to take employees at Tết (the Lunar New Year) to go visit relatives in the lowlands, this trip would be the only occasion during the year when the workers could see old friends and family (Trung tâm Nghiên cứu Phụ nữ 1987).

How did life on the SFE affect new subjectivities? Many workers had mixed feelings about their time in SFEs when I interviewed them years later. One friend, who by the 2000s was working for an environmental NGO, related his story. Like most SFE employees, Ngọc was born in the lowlands in Hà Tĩnh province and moved to work in an SFE based on social relations and his family's economic situation. When he was a teenager in the late 1970s, he decided to join his brother, who had been recruited for an SFE in Đắk Lắk province immediately after the cessation of war. Their family in Hà Tĩnh had little land, as his father was a widower who struggled to support his children. Once in the Central Highlands, rather than work directly for the SFE as a manual laborer like his brother, Ngọc decided to attend university, supported by his brother's wages, and get a forestry-engineering bachelor's degree. Once he completed this degree, Ngọc was hired by a local SFE in a district of Đắk Lắk. At that time, the SFE was quite remote, and it was surrounded by nearby ethnic Ede communities who were openly hostile to many of the SFE activities. Looking back from 2001, when I spoke with him, Ngọc related, "It was so difficult! I was so far away, and we didn't have anything to do, except maybe drink beer at night! And I got malaria several times, and it was really tough. But now that I look back on it, it was in some ways the best part of my life." The friendships he made with other workers at the SFE stayed with him his whole life, and his time at the SFE instilled in him a passion for rural development work and forestry. His skills were soon recognized by higher ups at the SFE, who recommended him for a promotion to the Forest Protection Department's provincial office. At that time, Vietnam had just started opening up more to foreign development organizations, and Ngọc decided to start learning English in 1990. These skills put him in the forefront of many of his colleagues, and he was accepted as staff for a German development organization in 1996 working on sustainable forest management projects, later traveling to Germany for a master's degree in forestry. He continues to stay in the field today, running an NGO on forest policy issues.[23]

For Ngọc, the SFE was the start of a lifelong career in the environment field, and, in this, we can say there were cases of "environmentality," or coming to care about the environment, through SFE employment. There were certainly others like Ngọc, and many forestry or environmental policy conferences in Vietnam today often turn into reunions of people who once worked together at SFEs. The exoticism of the SFEs was a draw for many ambitious young people who had never seen forests or been outside their home villages. Arriving in one of the remote SFEs after several days of travel must have been an exhilarating experience for some, as terrifying as it may have been for others. Some SFEs used elephants for skidding logs out of the woods, and one can only imagine the first sight of these creatures for young lowland peasants as they arrived in their new workplaces. For other workers, however, life on the SFE was a job, one they hoped would be only temporary, and it was a choice when not many other options were available. As difficult as forest work was, for many young Vietnamese coming of age in the 1960s and 1970s, it was far better than uncertain work in the delta eking out a living on crowded agricultural plots.[24] In a survey of SFE workers in the mid-1980s, 35 percent said they liked the SFE work and it suited them, while the rest said they did not know or did not like the work, but had had no other choice for employment (Trung tâm Nghiên cứu Phụ nữ 1987). This was particularly the case for young women from poorer families whose age cohort coincided with the devastating consequences of the Vietnam War, in which over one million North Vietnamese combatants and civilians were killed. For many of these women, the war casualties meant they would be unlikely to find a husband, and, in essence, their conjugal partner became the state in the form of employment at an SFE or state farm.[25] Nonetheless, despite making up the backbone of workers at SFEs from the 1960s to the 1980s, women did not find an egalitarian paradise in their new jobs, as they were largely excluded from management offices of SFEs. Not a single woman ever served as head of an SFE, although more than four hundred SFEs were eventually set up.[26]

INTERVENTIONS: ETHNIC MINORITIES, SWIDDEN AGRICULTURE AND RESETTLEMENT

In addition to the SFEs that transformed forest management in North Vietnam, another technology of environmental rule deployed by the DRV was the resettlement and sedentarization of ethnic minorities. Like the SFEs, these resettlement areas were often identified as sites for environmental management, but, in reality, they were aimed at the reshaping of the local indigenous peoples who dominated the mountainous areas. Like the French before them, DRV administrators were concerned with both the social and the environmental issues they associated with ethnic minorities, particularly the practice of swidden agriculture. But the DRV pursued a different policy strategy than the French, who had focused mostly on rezoning forests and trying to hold villages responsible for fires. The DRV, after initially trying to make peace with swidden agriculture as a way to keep "solidarity" with ethnic groups, became increasingly aggressive in its approach, eventually formulating plans to resettle and sedentarize in fixed villages a large number of minority citizens.

From the earliest days of the Indochinese Communist Party in the 1930s, reference was made to the importance of respecting ethnic minorities' ideas and culture, particularly since their support was needed for the anticolonial resistance. These promises to minorities were made tangible by the recognition of two autonomous zones (*khu tự trị*) soon after the founding of the DRV; these areas were given their own zonal assemblies, administrative committees, and militia forces (Jackson 1969; Kahin 1972). This conciliatory approach extended to the environmental practices of minorities as well, at least in the early years of the DRV. For example, a local report from the Black River area in 1955 reveals that local officials were working to guide production in new cooperatives and advising locals in the best methods for swiddens.[27]

By the late 1950s however, officials were still openly tolerating swidden but were becoming increasingly interested in regulating the practice more tightly, as officials realized that swidden would compete with the new SFEs for timber and wood, and they were not truly convinced it was an effective means of agricultural production, particularly for the large-scale cooperatives on the planning horizon at this time (Bộ Nông lâm 1957). Extensive swidden, with families spread out over large areas, planting one crop per year with limited technology, did not fit their socialist planners' visions. The Central Party Committee first officially addressed the idea that minorities might need to be resettled if they refused to give up swidden in Resolution 71/TW, dated March 23, 1963, which stated that it was the Party and State's business "to properly deal with the production and technical orientations aimed at gradually organizing resettlement and sedentarization, based on the principle of voluntarism, aiming at stabilizing and improving lives for people who practice shifting residence and shifting cultivation fields, in order to reduce forest burning and land degradation." In August of 1963, a group of central Party cadres including Ministry of Forestry officials visited the Tây Bắc Autonomous Zone to assess the problem. They reported back from their mission:

> Forest fires are very serious; for some years now the Forest Department has educated people to protect the forests but it has been limited in effect because of the habits of minorities in upland areas to burn forests to make swiddens, thus the habits of shifting cultivation of minorities in upland areas are not yet guided appropriately, and the situation of bare hills eating

> into green forests is each day more widespread. There are shortages of firewood in commune centers. The movement to replant forests has not yet been widespread. Every forest cadre has to be a propagandist, to educate people against swidden.[28]

These reports were increasingly accompanied by publications in academic journals and elsewhere, urging ethnic minorities to give up backwards and superstitious practices, and to progress to socialist economic development through settled agriculture and the application of technology, rather than through the pell-mell practices of swidden.[29]

Eventually the DRV tried the resettlement of swiddeners. Ho Chi Minh himself was alleged to have said: "The work of mobilizing sedentarization is one of the important jobs of our Party and State. If we do this job well it will contribute to the good implementation of our Party's minorities policy, contribute to the economic construction and culture of the mountains, and also contribute to the strengthening of our national defense" (Trường Chinh 1983, 13). While these actions had been discussed under the French, financial difficulties had prevented any large-scale sedentarization projects. The DRV decided to implement a Fixed Cultivation and Sedentarization Program (FCSP) (*Định canh Định cư*) starting in 1968. The FCSP's founding documents stated "the objectives of the program are both economic and social, i.e., build a stable production basis, to enable ethnic minorities to get work and participate in economic development, the building of newly developed rural areas, eradication of hunger, reduction of poverty, increased knowledge through education, better health, and arresting forest destruction."[30] The explicit link to forestry work was made by placing the FCSP office within the General Directorate of Forestry. Yet tellingly, FCSP target areas were not designated through ecological assessments of forests that needed conserving; rather, minority groups were prioritized by the percentage of households considered "settled," ranging from 100 percent of the Sán Chỉ to not a single member of the Si La and Chơ Ro ethnicities (SRV 2001). Yet what these terms of "settlement" referred to were not always clear, as many minorities did not practice itinerant shifting cultivation and already had fixed fields and homes.[31] On multiple occasions, government officials in provincial offices described swiddening to me not in terms of agricultural productivity or environmental sustainability, but on very subjective criteria of "under-development" (*không phát triển được*), being "non-modern" (*chưa hiện đại*), and "very wasteful of land" (*phí đất quá*).

The sedentarization projects were similar to the migration projects for

Kinh as both usually relied on voluntary movement of citizens into new areas, although there was considerable pressure—called "mobilization" (*vân động*)—to get people to move. In the case of the FCSP, officials encouraged households to move from higher areas to the lowlands, a process known as "coming down from the mountains" (*xuống núi* or *hạ sơn*). Households were often visited multiple times by state cadres and asked to make the move to new sites. Women and elderly people were the least likely to want to move, while young and newly formed households were more likely to heed the call in the hopes of acquiring land (a similar phenomenon to the Kinh migration programs). Unlike Kinh, who often moved hundreds of kilometers away, the FCSP usually moved people only a short distance of less than fifty kilometers to more accessible sites nears roads and valleys. Yet not all action was truly voluntary. Some households were pressured to leave by restricting their land use options and limiting the clearing of new upland fields, particularly where households were in conflict with newly formed SFEs.[32] In some cases, the SFE authorities themselves carried out the FCSP project goals by building houses and infrastructure for minorities, although employment in SFEs was almost never offered.

Soon after the FCSP began to be implemented, academic reports and newspaper articles started appearing, praising the sedentarization initiatives and the new subjectivities they engendered, specifically a stronger love of Party and socialism. One report on Yên Bái province noted prior to FCSP, minorities were "living by slash and burn, self-sufficiency and autarky, often succumbing to diseases, threatened by natural disasters, and living insecurely; the basis of the Party and the revolutionary masses was very weak, and local reactionaries looked for any way to cause difficulties for local revolutionary governments" (Trần Lộc 1971, 47). The minorities who had "come down" were said to have seen clearly the improvements in their lives, swidden areas were reduced by 75 percent, and the local minorities were now said to be "patriotic, to love the socialist way, and endlessly supportive of the army" (ibid., 48).

Not surprisingly, discussions with villagers who had been resettled under the FCSP had largely different remembrances of this program than these official success stories. The major complaint was there had been little investment in the new sites, and, as a result, there were often movements back into old villages even after being "sedentarized." The new resettlement sites were supposed to be provided with infrastructure, particularly houses and schools, while the movees would be expected to contribute labor to clearing wet rice fields. However, funds were low, and, in many places, resettlement proceeded with virtually no state investment whatsoever. One elderly man of Vân Kiều ethnicity, Mr. Núi [Mr. Mountain], who had been resettled in 1976 in Quảng Trị province by the FCSP, related the process:

> The new district authorities came to our village and said we needed to move. We were promised a better life with wet rice fields, but in fact there was no investment. Still, everyone moved. People complained a lot, but in the end, they had to accept it. In fact, the land here is much worse, and some people even lost wet rice fields at the other village site. In the 1990s, things got better because we could do lots of logging that we could sell, but once this area turned into a Nature Reserve [the Đakkrông Nature Reserve] all that stopped. The old site had land you could work for one year and it would feed you for two years. Here you work one year and there is still not enough to eat.[33]

Mr. Mountain's assertion that the movement had actually worsened their living conditions, and that the move had not been based on an assessment of land use in their old village, seems to bear out that resettlement has often been more about bringing people into the state's gaze, rather than any sort of environmental protection. Despite being located in the General Directorate of Forestry, and often being described as a program to reduce deforestation caused by swidden, the environmental objectives of FCSP have been the least emphasized and least funded part of the program. Indeed, the movement of people often required the clearing of even more land so new houses could be built (which required timber) and fields laid out (which required cutting and burning). In some cases, households agreed to move to a new residential site, while continuing to use swidden fields at their old location, which was the case for the community photographed in figure 2.01. At this site in Lào Cai province built for a Hmong community in 1995, households were using forest land near the new village as extra swidden fields, and some households were even returning to their former commune to use old swidden fields while maintaining a primary residence in the resettled state-built village. The living standard continued to be low, with some households experiencing a two to three month food shortage.

Compared to the SFE programs bringing thousands of Kinh to the highlands for forestry jobs, the FCSP was much less successful, both at transforming citizens and the environment. While no doubt many workers recruited to SFEs did not last long in the mountains and returned to their home villages, no SFEs failed due to lack of labor; someone could always be

FIGURE 2.01 Example of a Fixed Cultivation and Settlement Program resettlement site built for a Hmong ethnic community in Lào Cai province, 1996. Photo by author.

found to take an SFE job. Resettlement villages, however, were abandoned at a much higher rate. According to a 1990 review of the first twenty-two years of sedentarization policy, 2.8 million people in nearly 500,000 households had been targeted in 26 provinces. Of these, 1.9 million had actually been reached by the FCSP program in some form, of which 30 percent had shown good results and were "settled" and able to produce sufficient food. Of the remaining targeted households, 40 percent had only average results (meaning they had moved, but were still unable to produce sufficient food and still relied on old swidden fields or state subsidies), and 30 percent showed only weak results and were still "unsettled" (likely meaning they had completely abandoned the resettlement sites) (MOF 1990).

POSTWAR ENVIRONMENTAL AND SOCIAL CRISES

With the capture of Saigon by the North Vietnamese military on April 30, 1975, the US-Vietnam war was at an end, and a new reunified state, the Socialist Republic of Vietnam (SRV) emerged. Nationwide, the postwar problems were enormous and urgent, and included the need to prevent mass starvation by jumpstarting agriculture on abandoned and damaged lands, to resettle people displaced by war, and to address the imbalance in food production between North and South (Desbarats 1987). Forestry too was important, as a means to create employment in socialist enterprises in the former capitalist South, as well as to generate hard currency by exporting timber abroad.

While a great deal of attention in the West regarding the Vietnam War has focused on the environmental impacts of the deliberate destruction of forests in the former South Vietnam, including the spraying of chemical defoliants (Zierler 2011) (see Box 2.1), surprisingly, wartime damage of the South's forests was not the largest concern of the new SRV forest administration. The archives of the Ministry of Forestry show a complete absence of discussion about the effects of chemical defoliants for much of the late 1970s; rather, there were more concerns about such issues as shrapnel in trees that might affect timber processing, which was estimated to affect perhaps half the forests in South Vietnam (Lankester 1977). In perspective, war-caused environmental damages in the south appear to have been significantly less in terms of area than the environmental changes instigated by SFEs in the north.[34] The deforestation rates between North Vietnam and South Vietnam during the war years provide a useful point of comparison. From 1961 to 1975, South Vietnam's forest cover was estimated to have declined by around 140,000 hectares per year (von Meyenfeldt et al. 1978). Yet, from 1958 to 1974, in North Vietnam, an estimated 4,031,000 hectares of forests were lost, a decline of around 250,000 hectares a year (Nguyễn Cúc Sinh and Phan Lâm Lương 1986). Thus the DRV experienced nearly *twice* the deforestation, despite the defoliation campaigns confined to the South during the war.

Therefore it is perhaps understandable that the new forest administrators who were dispatched to the South after 1976 did not see defoliation and war damage; rather, they saw huge new areas of potential wood production, which would lead to employment for many that were impoverished after the war, and they were eager to use the North's model for forest exploitation, the SFE, as soon as possible, especially in the forest-rich provinces of the Central Highlands (Ngô Đình Thọ et al. 2006). New knowledge classifications were important to this effort. In July 1975, in order to rush the work of identifying the forest areas of the new reunified nation, the Prime Minister issued Decision 278, which classified all land with a slope of more than 25 percent as "forest land" for the exclusive goal of either tree planting or forest production and protection, a designation that holds to this day (MOF 1991b). Only six months after the fall of Saigon, the Forest Service had obtained aerial photos from the Ministry of Defense and was using them to assess the possibilities for SFEs in the South by mapping out large timber areas.[35] It was estimated the reunified country now had 8.2 million hectares of forests, of which only 2.5 million were located in the former North.[36]

The rich timber inventories of new areas were not the only reason why SFEs were set up in the South so rapidly after the close of the war; I was told

Box 2.1. Agent Orange and defoliation in South Vietnam

Defoliant attacks were carried out on approximately 2 million hectares of South Vietnam (primarily forests, but also including some agricultural lands) and another 200,000 to 300,000 hectares were likely damaged by napalm, bombing, and massive "Rome Plows" which flattened whole landscapes like a iron (Lankester 1977; Kemf 1988). The US military developed these strategies to combat the use of forests as cover for the enemy, and President John F. Kennedy personally gave the approval for the start of what would come to be known as Operation Ranch Hand, which sprayed over 20 million gallons of chemical defoliants, including Agent Orange, on areas of South Vietnam and Laos from the early 1960s until 1971 (Stellman et al. 2003). Studies initiated by the United States prior to the conclusion of the war indicated herbicidal sprayings had a very detrimental affect on forest structure, particularly in the short term, and resulted in considerable economic loss of merchantable timber. Trees usually lost leaves several weeks after spraying and remained bare for several months, and some more sensitive species died, particularly if they received repeated follow up spraying. One spraying was estimated to kill one out of every eight to ten trees; repeated applications raised the mortality rate to over 50 percent (Westing 1971).

The longer-term environmental impacts of the defoliation campaigns have been less well understood, as "the degree of initial damage and subsequent rate of recovery, however, depend on many factors, including the species of vegetation, herbicide concentration, total area affected (smaller areas will usually recolonize much faster), the terrain, soil fertility, soil microorganism, and weather" (Hay 1983, 208). In some areas forests were able to regenerate fairly quickly, although the exact forest composition favored secondary growth species (Ashton 1986). Multiply sprayed forest areas fared the worst, and, where tall canopy trees were killed, invasive grasses like *Imperata* and *Pennisofolum* (known as "American grass") prevented natural forest generation. The Mã Đà forest of Đồng Nai province and A Lưới valley of Thừa Thiên Huế province are two areas where the footprint of such defoliation is most obvious (Phùng Tửu Bôi pers. comm. 2002). Mangrove forests were also particularly sensitive to the herbicides; such stands tended to be dominated by plants of one species and of a similar age, and, consequently, a herbicide application tended to affect the individual trees in a uniform manner. It was estimated around 150,000 hectares of mangrove and rear mangrove (*Melaleuca*) were completely deforested during the war (Lankester 1977). Many of these mangroves did not grow back naturally, and had to be mechanically replanted.

by an SFE director in Thừa Thiên Huế province who had been there for many years that one major reason for SFEs was to provide security against counter-revolutionary forces still operating in the postwar period, and in particular, against FULRO (*Front Unifié de Lutte des Races Oprimées*, the "United Front for the Struggle of the Oppressed Races," an ethnic minority guerrilla force in the Central Highlands).[37] The FCSP program was also brought to the former South once the country was reunited given this need for surveillance and control of potentially dangerous citizens (Salemink 2003). Particularly in the Central Highlands, minority villages were quickly targeted for FCSP interventions to stop swidden and "settle" populations (Lưu Hùng 1986). But unlike in the North, one additional factor that encouraged the use of FCSP projects as a tool of intervention was the practice of communal longhouse living of some ethnic groups in the former South. FCSP resettlements were used to break up these longhouses and move communities into individual pre-built houses in attempts to change the cultural allegiance of minorities from their clans and relatives to the new reunified state. Policies were also adopted to abolish traditional customary practices, such as funeral rites, buffalo sacrifices, and harvest feasts associated with swidden. These practices were seen as "wasteful" (*lãng phí*) and the association with spirits was "superstitious" (*mê tín*) (Nông Quốc Chấn 1978). Thus environmental excuses regarding the damages of swidden were used to justify what were most decidedly *cultural* policies.

New FCSP villages followed fairly standard forms and conventions: households were to live Kinh-style in individual houses, farm wet-rice agriculture, and declare their loyalty to the Party and the state. Some FCSP villages went so far as to affix rules on conducting conduct on signboards at the entrance; in the settlement site pictured in figure 2.02, these rules included producing food through "active labor" and "progressive use of science and technology" while not using swidden or deforesting; preserving the peace and security; and "throwing away" old superstitions, customs and habits. These sorts of policies resulted in rapid social and cultural change, far more so than environmental change (McElwee 2004).

Supplementing FCSP projects in the former South were the creation of extensive new SFEs, and to bolster these areas and secure them for the state, workers were mobilized from the North; for example, four hundred cadres from Hương Sơn SFE in Hà Tĩnh were sent to Đắk Lắk province to build SFEs in 1980 (Văn Ngọc Thành 2005). Many other state farms, such as for coffee, also spread in the former South, particularly in the Central Highlands. Unlike in the previous wartime period, many of the postwar recruits to SFEs

FIGURE 2.02 Posted rules for conduct in a Fixed Cultivation and Settlement Program site in Đồng Nai province, 2014. Photo by author.

and state farms were men, including army veterans and youth pioneers (*Thanh Niên Xung Phong*), who had served in highland areas during the war. One informant who worked for the Forest Inventory and Planning Institute was sent south to inventory areas for SFEs in Đồng Nai province, and stated that in many places, SFE administration and infrastructure actually preceded other forms of state-building, such as the establishment of commune (*xã*) administration (eg, People's Committees and police departments).[38] For many years, an SFE was the only presence of the new state in remote areas of the former South.

The new SFEs were tasked with meeting the enormous demand for timber for postwar reconstruction material; it was estimated ten million homes needed to be reconstructed (Kemf 1988). Wood was also needed to rebuild the Hanoi to Saigon railroad, out of operation for many years during the war.[39] The SFEs in the North needed to be refurbished as well; many were missing equipment, such as vehicles, skidders, and sawmills that had been bombed or pressed into war service (Văn Ngọc Thành 2005). The first unified Five Year Plan for the SRV (1976–80) ambitiously called for putting 4.3 million hectares under SFE control, or around one-third of the total land area of the reunified country; former areas of South Vietnam, such as the Central Highlands and U Minh mangrove and swamp forests of the Mekong Delta, were particularly targeted (Nguyễn Tiến Hưng 1989). Timber production rapidly expanded, reaching a peak of over 1.7 million m^3/yr in 1977, the highest it had ever been (GSO 1992). Even this rapid rate of cutting was not as high as officials wanted; a directive laid out plans to exploit 2 million m^3 in 1978, moving to 3.5 million m^3 by 1980.[40] Much of this expanded production was aimed at export markets in order to earn foreign currency in the lean postwar years. The issue was considered so important the state vice-president directly convened meetings in the Central Highlands on increasing timber exports, which during the 1980s went mostly to the Soviet Union and Hong Kong.[41]

In order to meet these demands, SFEs became larger, both in terms of area and labor, than they had ever been in North Vietnam. A clear example of "new-style" SFEs can be seen in the Ea Súp Forestry-Agriculture-Industry Union, which was created in Đắk Lắk province in 1979. Prior to establishment of Ea Súp, there were approximately 10,000 local people living in the area, primarily of Ede ethnicity. Ea Súp SFE recruited more than 20,000 lowlanders to take jobs, and the hiring was so rapid that many workers had few to no skills or experience. The SFE controlled more than 300,000 hectares of deciduous forest, and the vast majority of lands were scheduled for production of timber, veneer wood, and pulp, which were supposed to follow a twenty-year cutting rotation, with replacement of trees by aggressive reforestation. Yet these targets were quickly missed; in the first ten years of the SFE, half the standing volume of timber originally surveyed had already been removed (over 1 million m^3), much of it for export, and little had been replanted (Phạm Hữu Văn 1989).

These problems with shortened cutting cycles leading to unsustainable rates of logging occurred in many other SFEs, where forest management prescriptions were altered in order to achieve the high production targets imposed on them by the first Five Year Plan. The selection felling cycles, which were set at twenty to thirty years in the 1960s, were reduced to five to ten years by the late 1970s, and minimum felling diameters shrank from 50 to 30 cm (Ogle et al. 1998). Another challenge was the fixity of market prices; as a socialist country, free market sales of timber were to be forbidden, but the black market thrived and created strong demand.[42] Other bureaucratic hurdles took up enormous amounts of time for SFE workers, such as trying to procure vehicles, gas, and even stamps in a nonmarket system. Every request for any major item needed by the SFE had to go to either the provincial or central government for budget allocations, and the archival dossiers of the General Directorate of Forestry in the late 1970s and 1980s are filled with the minutiae of socialist planning—requests for retirement for injured workers, for moving residency papers, for bands for sawmills, or gas for vehicles.[43] Some SFEs chose to "'go outside' the planned economy in order to obtain materials, while the latter compelled them to provide other sources of income for their workers in order to shore up their living standards, providing both the necessity and the opportunity for illegal timber-getting" (Beresford and Fraser 1992, 13). SFEs were supposed to give a fixed percentage of their profits toward the local state budget, but many times kept a substantial proportion of this for themselves and under-reported timber exploited beyond projected targets.[44]

Ecologically, little attention was paid to noneconomic values of forests in the postwar era: ecology took a back seat to industrial socialism. Books and research works from the Forest Science Institute and other organizations from 1975–86 primarily focused on research on industrial trees and on expanding fast-growing timber supplies. Such inattention appeared to be in inverse proportion to what was needed, as increasingly damaging ecological impacts were evident in many areas. For example, the poor equipment at SFEs meant many workers would "gully" their logs by throwing them down hills rather than manually skidding them out, leading to erosion and more forest loss (Ogle et al. 1998). And although SFEs were always supposed to have reforestation plans prepared to restock forests, in reality, they all exploited far more than they planted. The 1976 Five Year Plan called for reforestation of 1.2 million hectares, but only 8 percent of this target was achieved (Agarwal 1984).

Socially, too, many SFEs fell into crisis in the postwar era. By the 1980s, there were over four hundred SFEs—around 76 under the central Ministry of Forestry, 199 under Provincial People's Committees, and 138 under District People's Committees—controlling huge swaths of land in much of the rural countryside (MARD 2001b). These SFEs were not evenly located, with some provinces having twenty or more SFEs and up to 70 percent of their land area managed by SFEs, while other provinces had few to none (see map 2.03). Local populations often protested against the nationalization of their forests by SFEs with acts of sabotage, such as arson or expansion of agricultural land into SFE forests, escalating up to assaults on SFE personnel. Arson had long been a nemesis of local SFEs, as it was one of the most common forms of resistance and rural protest.[45] SFEs also ran into conflicts with nearby cooperatives, as the need for buffalo to work the expanding rice fields during agricultural collectivization increased demands for pastures. One retired SFE

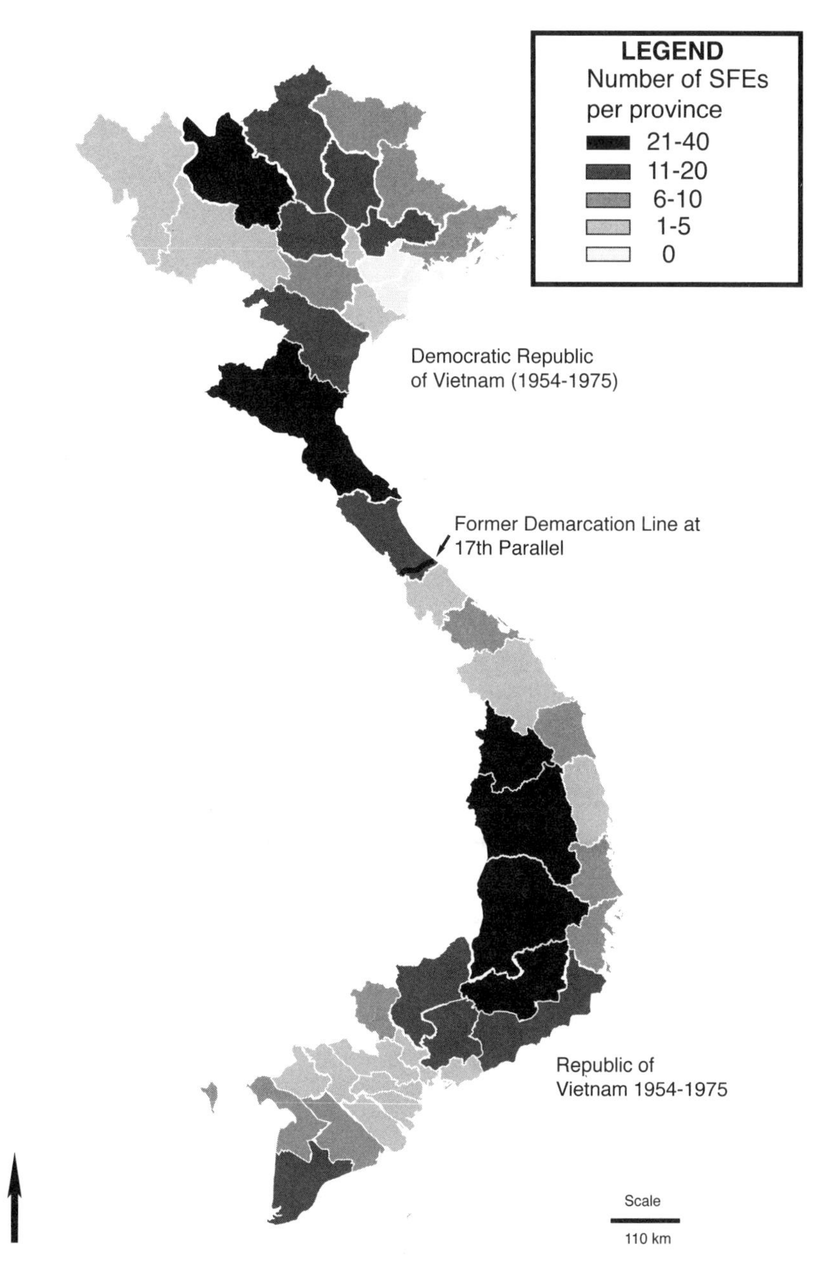

MAP 2.03 Distribution of State Forest Enterprises in reunified SRV circa 1986. Note provincial boundaries in the late 1980s are different from today. Data from MOF 1991a. Base map from Vietnam location map by Uwe Dedering on Wikimedia Commons. Map by author.

cadre in Hà Tĩnh blamed the majority of "bare hills" surrounding his SFE on the practice of letting fires spread widely to provide green pasturage in the 1960s and 1970s, rather than forest cutting.

Sabotage against the SFEs could also be internal. As timber receipts began to slow down after the overcutting of the late 1970s, payment of workers salaries was often tardy, sometimes as much as eight to nine months late; a survey in 1987 listed late salary payments as the number one complaint of SFE employees (Trung tâm Nghiên cứu Phụ nữ 1987). Salaries also did not keep up with the rampant inflation in the postwar era; estimating salaries in terms of rice equivalent, in 1987 an SFE worker could buy 117 kg of rice, but by 1994 their salaries only paid for 30 kg (Nghiêm Thị Yến 1994). As a result, the average number of working days had dropped to only seventeen per month, and many of these were not full days, as workers would do desultory work in the morning, then return in the afternoon to private agricultural plots (Trung tâm Nghiên cứu Phụ nữ 1987). SFE employees increasingly focused more on home gardens than on their jobs, as state rice rations were rarely sufficient, and there is strong evidence that over-clearance and lack of attention to prescribed forest-cutting cycles were in fact deliberate attempts to keep forest land clear of trees so employees could expand household agricultural plots. By the 1980s, workers were also no longer satisfied living in collective housing; less than half the workers still used the provided dorms, while the rest cut down trees to make their own private houses. Further, SFE workers confessed to collecting other forest products illicitly to sell on the black market for extra money to buy necessities in the cooperative stores.[46] These acts of resistance were a dramatic turnaround from the solidarity and comradeship at SFEs so vaunted in the 1960s during wartime, and indicated the new socialist subjects the state had hoped to produce from environmental rule were not accepting of their changed economic conditions.

CONCLUSION

The metamorphosis in social relations, settlement patterns, and environmental practices in the aftermath of war and socialism was profound. The French had treated forest reserves as a place to use state authority to extract tax revenues, while under the DRV forest management was about labor and state-led development, and secondarily about forests. Yet what is striking is how similar some of the projects for environmental rule implemented by the DRV were to those of the previous French administration. The French colonial regime and the DRV shared a hostility to swidden agriculturalists, a top-down forest bureaucracy, and a focus on reorganization of the forest estate through the replacement of diverse natural forests with more economically viable plantations. Indeed, the DRV ended up reproducing many of the same errors as the French, including the limited focus on a small number of hardwood species, a pattern of over-harvesting inappropriate to the ecology of heterogeneous forests, and ineffective approaches to swidden.

To be sure, there were significant differences in the two eras' approaches to new sources of authority and new regimes of practice to carry out environmental rule. SFEs were a turnaround from the laissez-faire management of the French forest service, which had relied on concessionaires to monitor themselves. The sedentarization sites applied to millions of ethnic minorities fundamentally reconfigured upland settlement patterns in ways the French could never have imagined. And the new networks of rule in the DRV influencing forest management included advisors from the Soviet bloc and Eastern Europe who encouraged the industrialization of the forest sector, leading to an emphasis on top-down SFEs at the expense of local and decentralized forest management. Ultimately, the DRV's nationalization of forests exerted a level of control over land and people unmatched by earlier eras, and the new subjectivities produced by forest labor and resettlements resulted in profoundly different lives for thousands of people who took part in these projects.

The new technologies of rule devised for forest management, including both SFEs and resettlement sites, required certain kinds of problematizations (such as the idea that swidden agriculture was worse for forest structure than large-scale logging) and certain kinds of knowledge (such as the classification of forests by degree of human disturbance), while other types of problems and knowledge were obscured (such as local knowledge about tree management or existing norms on collective forest use). These new forms of rule also relied on novel networks of circulation, both of people and ideas. Migrants to SFEs and other work in the uplands often recruited each other, bringing relatives up to join them in their new lives, as Ngọc's brother did. The voluntary nature of movement stressed freedom as well as reinvention, but the disciplinary aspects of SFEs and sedentarization sites were also hard to ignore. The SFEs and the resettlement sites both served as objects of environmental rule that could be circulated into new areas when the SRV was unified, with little to no change, despite the difference in ecologies and histories between north and south at the close of the Vietnam War.

Despite the use of environmental rhetoric about sustainable forest production or the devastation to ecology caused by swidden, SFEs and sedentarization sites were resolutely about people, not about nature. For SFE employees, to be a new socialist citizen was to take part, however ambivalently, in new forms of family and labor in unfamiliar highland areas. For sedentarized ethnic minorities, to be a new "settled" citizen was to be in an assemblage of multiple factors, some spatial (to not be living on high slopes, to be in a concentrated settlement near government officials), some economic (to be planting wet rice, to be producing sufficient food to sell as well as consume), and some social (to have given up superstitions and to be "civilized").What is truly remarkable, however, is how hidden these histories have largely remained; despite much attention to agricultural cooperativization during the DRV era (e.g., Kerkvliet 2005; MacLean 2013), there is a virtual silence about the role of the SFE in similar processes of land territorialization and citizen subject formation.[47]

The most obvious long-term impact of this era of environmental rule has been the uneven distribution of land rights between forests and agriculture, between the center and localities, and between Kinh and minorities, as these government programs of migration and resettlement have now completely changed the spatial and ethnic composition of highland areas. Later difficulties in reestablishing or inventing anew land and resource management practices after decollectivization in the 1990s were not surprising, as residents of the DRV had been told to discard outdated ways of managing forests and food production. Old practices, such as swidden agriculture or community tenure, were to be thrown out and new persons remade: into socialist workers and settled assimilated citizens. Knowledge of forests that was local and specific was to be eliminated and replaced with industrial management for large-scale timber production. Locally devised rules for forest management were to be dismantled in favor of state-led development, so the benefits of forestry for the entire nation could be shared. The perception among local peoples, however, was that this "sharing" largely amounted to "stealing" of forest resources by the state, and clashes over what the state decreed and what people actually did on the ground is the focus of the remaining chapters of this book.

3 Illegal Loggers and Heroic Rangers

The Discovery of Deforestation in Đổi Mới (Renovation) Vietnam

FROM 1986 TO 1992, VIETNAM WAS ESTIMATED TO BE LOSING about 200,000 hectares of forest a year, giving the country a deforestation rate of 2.4 percent, and the UN Food and Agriculture Organization (FAO) ranked Vietnam as having the second highest rate of deforestation in the world in the late 1990s (Veillieux 1994; FAO 2005). Domestically, stories of illegal logging dominated the newspapers with rising frequency, often focusing on the phenomenon of the *lâm tặc*, or illegal logger, a person who deforested with impunity, usually because he had connections. It was not a coincidence that this emergence of the "problem" of deforestation occurred at approximately the same time as Vietnam began to move away from state socialism, a process known as renovation (*đổi mới*: lit. "new changes"). Like the glasnost and perestroika era occurring at approximately the same time in the Soviet Union, the renovation period saw the dissolution of many aspects of the state-managed economy, and new questions about who had rights and authority over environmental problems subsequently surfaced. Fears of declining resources and a loss of state control over the forest sector, as indicated by concerns over excessive "illegal" logging and deforestation, led to new approaches to forest management and environmental rule at this time.

Different policy responses to deforestation corresponded to who or what was problematized as the driver. Officials in the DRV had begun to recognize deforestation as an issue in the 1970s, although to them, the concept of "deforestation" primarily meant "non-state forest use." It was not until *đổi mới* that deforestation truly became a widespread societal problem. Deforestation became a matter of concern through multiple channels, including the first access to remotely sensed data and satellite technologies, expanding media reports, and increasing attention from international donors. Differ-t networks came to have knowledge about deforestation as information circulated, new stakeholders were enrolled in ideas regarding how to combat deforestation, and some actors were blamed for deforestation, while others were not. Ultimately, how these multiple discourses about deforestation were made (and by whom), which discourses were heard (and which ignored), and how these discourses were deployed (and when), all served to influence the policies and interventions that were imposed upon forests in the post–*đổi mới* era.

New forest interventions that emerged after 1986 included novel ways to classify and map forests, and new ways of thinking about their management, including a move away from timber production and towards other values for forests, such as biodiversity conservation or aesthetic enjoyment by tourists. These reclassification projects primarily treated deforestation as a technical problem that resulted from a lack of clarity regarding land use zoning and led to respatialized planning of forests to more "rationally" allocate where different objectives (logging, watershed conservation, or biodiversity) might be best fostered. These activities were assisted by bilateral and multilateral institutions, which began to provide large amounts of aid to Vietnam, and the ailing forest sector reaped considerable benefits from this infusion of cash. Logging bans combined with more strictly protected areas grew rapidly in the 1990s, although many of them remained "paper parks" in practice. Strengthened law enforcement of forest laws was an additional focus at this time.

Yet these protected areas were not solely driven by environmental concerns. Some actors, particularly international conservation NGOs who began to work in Vietnam in the 1980s for the first time, focused on biodiversity conservation and the need to preserve "natural" forest ecosystems and undertook species inventories and mapping exercises to highlight where this biodiversity might be found. But many of these NGOs' recommendations on where to put new protected areas were not taken up by the state, and, instead, many "new" protected areas were placed at sites of previous SFEs. As part of this transformation, SFEs that were out of money and timber by the 1980s suddenly had value again, and SFE employees could keep their jobs and become "forest protectors." Thus the protected areas system that arose in Vietnam during *đổi mới* was driven as much by concerns about employment and continuing state control over land as it was about biodiversity, in a classic example of environmental rule.

The newfound concern over deforestation can also be seen as an outgrowth of concerns about the changing relationship between households and the state in an era of more limited government. Environmental rule became a tool to remake the relationship between the state and individual households over economic activities and land tenure as the government began to withdraw from cooperatives and other large-scale, state-socialist development projects. Whereas the previous postwar period had been primarily about labor and population in relation to forests, the *đổi mới* period can be characterized by a new focus on forests in relation to populations, with state attempts to impose clear dividing lines between "protected areas," "production areas," and people. The state would manage the former, and citizens would be asked to self-regulate their forest use by abiding by new laws and pledging to protect forests. This was a major turnaround from the previous socialist era of SFEs, which brought people into the forests, encouraged forest exploitation and use, and erased dividing lines between nature and society.

But the peasants dwelling near forests and the swiddening minorities, who were the targets of the interventions of a strengthened forest ranger service, new forest classification systems, and the criminalization of most non-state forest activities, largely ignored these policies and continued their patterns of forest use. This resistance was not met with coercion so much as it was met with attempts to change subjectivities: in the *đổi mới* era, remaking forest spaces did not involve the top-down compulsion that had characterized state resettlement, SFEs, and sedentarization sites in previous eras. A more pervasive change in the move to market liberalism was the co-optation of discourses of development, modernization, and conservation, which urged citizens to take on new roles and subjectivities as responsible land users.

As a result, new forest subjects emerged in the *đổi mới* era: a category of people known as "illegal loggers," *lâm tặc*, who were not following the new dictates to conserve protected forests; and state forest rangers (known as *Kiểm lâm*), who were on the front lines of regulating deforestation from the *lâm tặc*. The ranger was supposed to embody a new way of protecting forests, privileging the rule of law and enforcing new policies based on scientific classifications of forest types. In reality, forest rangers became contested objects of environmental rule and played ambivalent and contradictory roles. Rangers became objects, in that they were to be the transmitters of governmental authority over dwindling forests in a decentralizing era, but rangers were subjects as well, with human desires, concerns, and demands that influenced how effective environmental rule was on the ground. Further, the rapid commercialization that followed the opening of a market economy created high demand for timber for a burgeoning middle class, and logging in place in the early 1990s drove down the supply of logs, creating an higher appetite for luxury woods. Rangers were now positioned to be brokers of access to valuable timber, and in many places they became involved with activities that were in contravention of their stated roles; accusations of corruption and violence against rangers erupted throughout the 1990s and 2000s. By contrasting the roles and subjectivities of forest rangers with those of the figures they were supposed to be protecting against, we can see that, to many locals, the *lâm tặc* (illegal logger) and the *Kiểm lâm* (ranger) were one and the same person. This skepticism over whether the state and its representatives were genuinely interested in conservation, or in lining their own pockets, played an important role in thwarting environmental rule during this period.

ĐỔI MỚI AND THE EMERGENCE OF "DEFORESTATION"

The late 1970s and early 1980s were a difficult time in Vietnam. Rampant inflation and the breakdown of agricultural cooperatives dominated people's lives, with several years of successive droughts and floods leading to food shortages from 1976 to 1980 (Agarwal 1984). The forest sector was in crisis in the late 1970s and 1980s as well. The General Directorate of Forestry had been promoted to the Ministry of Forestry (MOF) in 1976, indicating this sector's importance in postwar economic construction. Over half of all provinces (thirty-eight) had at least one SFE in operation by 1990, and more than four hundred total SFEs across the nation occupied nearly 6 million hectares of land, with an average area of 15,854 hectares (Armitage 1990). Forestry labor hit a peak of employment around 1988, while 1979 was the high point of timber production in Vietnam, with around 1.9 million cubic meters of timber cut that year (GSO 1992, 247) (table 3.01). But it was clear to many that the SFEs could not keep this pace going, given the poor state of much of the remaining forests and the lack of success in reforesting previously logged lands (Durst et al. 2001). Many SFEs did not invest in road building or reforestation, and therefore simply kept logging the most accessible areas of forest over and over, preventing them from recovering or regenerating (Armitage 1990).

The effects of the large-scale logging could no longer be ignored, and a concern began to develop among many officials that Vietnam's forests were being depleted at an alarming and irreversible rate. However, blame for the declining availability of timber still was not fully laid at the feet of SFEs, but rather on the neighboring populations that refused to recognize the SFE and

TABLE 3.01 Forest production in state forest enterprises in the postwar period

Year	Roundwood, in m3	Firewood, in 1,000 stere
1976	1,635,000	731
1977	1,659,000	1,045
1978	1,564,200	876
1979	1,740,300	930
1980	1,576,600	874
1981	1,354,900	1,203
1982	1,294,600	1,080
1983	1,401,700	1,022
1984	1,476,000	1,204
1985	1,439,000	1,089
1986	1,462,100	971
1987	1,498,600	875
1988	1,340,300	745
1989	988,000	506
1990	1,081,000	407

Source: GSO 1992, p. 247.

Forest Departments' authority over forest management. For example, in the province of Hà Tĩnh, authorities lamented that:

> Forests are declining in all places, caused by human activities. . . . The forest decline increased partially due to the destruction of the war, however, the decline is still mainly because of pressures from reckless forest exploitation (*khai thác rừng bừa bãi*).. . . 15,000 hectares of forest and forest land is lost each year; of that, one half the area lost is due to clearing fields (*khai hoang*) to make agricultural land, and one half is due to people sneaking into the forest storehouses to exploit fuelwood and timber. (Nguyễn Phúc Khánh and Nguyễn Văn Trọng 1986, 37)

Similarly, official concern over "deforestation" (*nạn phá rừng*) through the 1960s and 1970s was more about timber wastage than larger concerns about forest ecology or biodiversity. A Ministry of Agriculture report in the early 1970s, for example, noted that:

> The work of protecting the forests is still very weak. Widespread and serious incidents of deforestation are happening. Swiddens and forest fires each year consume tens of thousands of hectares of forest, shrinking the forest area seriously, reducing watershed areas, old growth forests, forests with high economic value, and many other areas are continuously deforested (*bị phá*). . . . With organization [like this] that is deficient, weak and thin, then the receipts from forest products are significantly lost.[1]

Such thinking primarily reflected economic concerns about loss of forest timber more than environmental concerns. Thus activities to combat deforestation were more about tightening state control and instituting regulations on forest cutting, forest fire protection, and prevention of swidden practices than on seriously reconsidering the ecological costs of over-lumbering.

In response to these challenges, an Ordinance on the Protection of Forests [*Pháp lệnh quy định việc bảo vệ rừng*] was promulgated by the National Assembly in fall of 1972. The ordinance reiterated the ownership of forests by the state, and explicitly stated that no forest exploitation was to take place without the approval of appropriate authorities. Swidden was expressly forbidden, and forest fires were criminalized. The General Directorate of Forestry emphasized the need to translate this law into minority languages, as they were some of the primary subjects at whom the law was directed.[2] But the major innovation of the Ordinance was the founding of a People's Forest Protection Service (*Kiểm lâm Nhân dân*, sometimes also translated in English as the "Forest Inspectorate" or the Forest Ranger Service) whose role would include the authority to enforce forest laws.[3]

Kiểm lâm (KL) officers were given the right to monitor and inspect all forests, including those managed by SFEs; to enforce tax and trade in forest products; and to determine violations of any national forest laws, even on non-state lands. Local KL branches reported to the provincial and district People's Committees as well as to the national General Directorate of Forestry, a form of both horizontal and vertical integration that was typical of Vietnamese bureaucracy but which often created overlapping expectations. An additional problem was that the KL had little to no experience in scientific management or nature preservation, and little funding was provided to help the KL do its work. The fact that deforestation continued to increase in scale and scope after the founding of the ranger service provides some

indication of their lack of effectiveness, and it was not until the 1990s th
the KL received more significant state financial and legal support.

VISUALIZING DEFORESTATION

The *đổi mới* era coincided with the ability to visualize forest change at the national level for the first time through satellite photos obtained from the Soviet Union. The aerial views revealed what one observer called

> a shocking state of affairs. In the province of Nghệ Tĩnh, which is the third most important forested province in the country, 28% of the rich, broad-leaved forests had been degraded or disappeared entirely between 1977 and 1981. Similar case studies showed that since 1976 another million hectares of natural forests in the country have been degraded, and in 1982 Vietnam had only some 7.3 million hectares of forest land. (Agarwal 1984, 12)

Maps created in 1990 by the Forest Inventory and Planning Institute, a division under the Ministry of Forestry, attempted to use satellite data and reports of forest decline from individual provinces to create visuals showing the significant nation-wide change in forest cover (figure 3.01). The most serious decline appears to have occurred between 1976 and 1990, and particularly from 1976 to 1985, during the first two Five Year Plans of the reunified SRV (see table 3.02).

However, one should take reports about exact figures of forests and forest loss with considerable caution, as many statistics about forest cover change in the 1940s to 1980s suffered from a lack of accuracy. Most national compilations were based on reported changes by local provinces, not on rigorous assessment of aerial and satellite imagery. An FAO review in 1990 noted that most local inventories in Vietnam were not accurate, and that the vast majority of estimates of forest land were overstatements (Armitage 1990). Attempts to assess deforestation were further complicated by the variable definitions of "forest" used by different ministries, officials, and localities (Sowerwine 2004). According to the Forest Resources Protection and Development Act of 1991, the first nationwide law that attempted to classify and codify where forest management could take place (discussed further below), "forest land (*đất rừng*) is defined as 'forested land' and 'non-forested lands for which plans have been made for forest plantation.'" No other definition was provided of what a "forest" was, therefore, the official government definition of "forest land" included land that did not necessarily have any trees on it. In a 1990 review of forest policy undertaken by donors, it was estimated that the

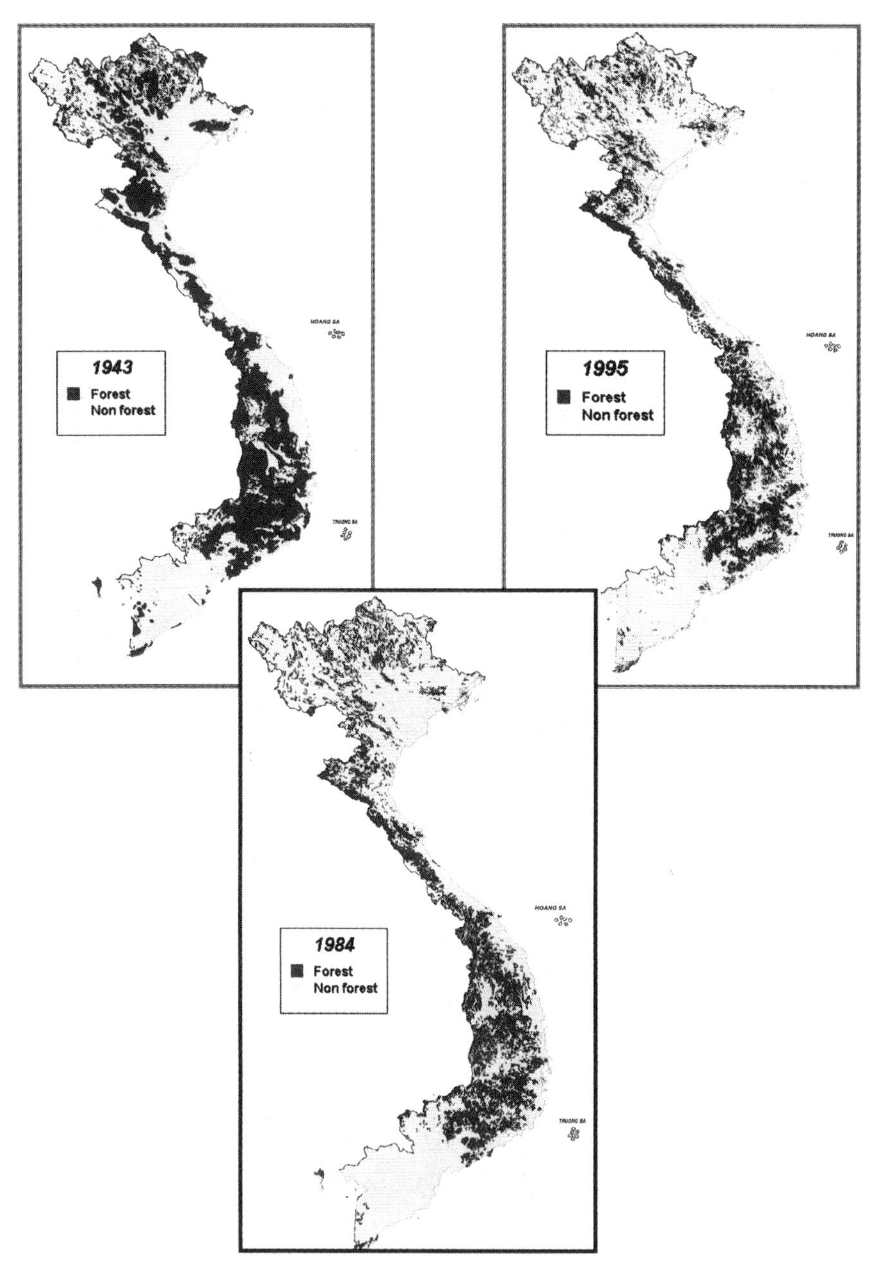

FIGURE 3.01 Map of deforestation trends, 1943–1990. Map from Wege et al. 1999, based on Ministry of Forestry data.

TABLE 3.02 Official figures on forest cover change in Vietnam 1943–1983 (in thousand ha)

Region	Year	Total Area	Forest Cover	Percentage Cover
Bắc Bộ (North)	1943	11,575	5,500	47.6
	1975	11,000	2,200	20
	1983	11,575	1,862	16.1
Trung Bộ (Center)	1943	14,760	6,000	40.6
	1975	15,860	6,215	39.2
	1983	15,232	5,244	34.4
Nam Bộ (South)	1943	6,470	2,000	30.9
	1975	6,040	1,085	17.7
	1983	6,335	704	11.1
Vietnam Total	1943	32,804	14,325	43.7
	1975	32,900	9,500	29.1
	1983	32,162	7,815	23.6

Source: MOF, 1991b, p. 33.

government had designated 19 million hectares as "forest land" (57 percent of the total country area), yet less than half, only 9 million ha, had any degree of tree cover (Armitage 1990).

This definition of "forest land" as "places the government has decided to call forest" meant that visual assessment or local practice of what was a forest, and what was a fallow field or pasture, were not reconcilable (Mellac 1998). Some donors who were involved in funding forest activity in Vietnam openly questioned the official classification system; in an interview with a foreign staff member of an international conservation NGO, the government's land plans were deemed "very dodgy," with the staffer adding, "what is forest anyway, the government plants a tree every kilometer and calls that a forest."[4] The ontological complexity of what a forest was could not be solved even with the increasing use of remotely sensed images; as many scholars have pointed out, these technologies can often increase confusion and contestation over forest definitions, not clarify them (Fairhead and Leach 1998; Robbins 2001b). For example, Meyfroidt and Lambin (2008a, 6) found that even using multiple remotely sensed data sets, estimations of total forest cover in Vietnam during the 1991–93 period ranged from 23.7 percent to 53.7 percent, depending on

TABLE 3.03 Definitions of "forest land" under control of the Cẩm Xuyên State Forest Enterprise, 2001

Type of Land	Total Ha	Percentage total lands
"Forest land" with tree cover	3,527 ha	34
Of that land:		
Forest land with natural forest	382 ha	11
Forest land with plantations	2,495 ha	71
Forest land with naturally regenerating forest	650 ha	18
"Forest land" without tree cover	5,522 ha	53
Other land area not considered forestry land (residential, water surface, etc)	1180 ha	13
Total Area of SFE	10,229 ha	100

Source: Interview with director of SFE, April 2001.

whose map was being used. One reason for the wide variation was the fact that satellite data and aerial surveys were not shared outside the government due to fears over national security, so accuracy was difficult to check.

At the same time as the official classifications for forests were being developed and used, local definitions too were being deployed, often as claims over land rights. Local peoples usually have alternative forest classification systems that differ significantly from national ones; for example, such local classifications might group forests by age and time, rather than by ecological types.[5] I was able to compare land descriptions and quality between a state-designated protected area, the Kẻ Gỗ Nature Reserve, an SFE (the Cẩm Xuyên SFE), and locally managed village common land during fieldwork in 2001. Local people had their own definitions of forest that differed markedly from those of officials and maps. Locals never called land without trees, for example, "forest land," referring to these areas as fields (*nương*), pastures (*bãi*), and fuelwood collecting areas (*khu vực khai thác củi*), depending on the end use. "Open forest" areas (*rừng thưa*) were lands with tree stems generally taller than people but without a dense canopy. "High forest" (*rừng rậm*) could be defined by taller trees, generally closing off the view of the sky, and undergrowth of a variety of shade-loving plants. While the official figures on land cover that would have been reported upwards from this district, and compiled in national statistics, would have included more than 10,000 hectares of state-designated "forest land," as table 3.03 shows, less than 400 hectares of this had naturally growing tree cover and would have been recognized by my local informants as a real forest.

PROBLEMATIZING DRIVERS OF DEFORESTATION

In the problematization that followed the discovery of deforestation in the 1970s and 1980s, certain culprits were blamed while others exonerated. The fact that there was no standard and agreed upon definition of deforestation or "forest land," nor were there systematic attempts to assess reasons for forest cover changes, meant that multiple narratives about the drivers of change circulated through different networks. Who was being blamed for deforestation usually said more about the position of the accuser than any accurate assessment of trends and statistics, because the latter simply were not being compiled or analyzed (De Koninck 1999).

For those working in the government sector, "deforestation" was nearly always an action that people did outside of the directives of the state. Drivers of these activities were often linked to poverty or overpopulation, not the presence of SFEs controlling large swaths of forests, which consequently limited others to the remaining poorer quality lands. For example, the head of the Forest Science Institute blamed "poor people...who are highly dependent on forests for food, fuel and income" (Vietnam News Service 1997, 3). A provincial forest director attributed forest loss to Malthusian population growth: "There is a common law of underdeveloped countries with rapid population growth. To get land for production and wood use, people have to rush to the forests to earn their living" (Kibel 1995, 242). Not surprisingly, given the state's long dislike of the practice, swidden agriculture also continued to be a frequent culprit. A Ministry of Forestry report from 1990 noted, "Among the main reason of deforestation is shifting cultivation, a farming practice used by minority ethnic [sic] in mountain areas. . . . The local people are aware that the loss of forest cover has many serious implications" (MOF 1990, 8). The vice-chairman of the government committee on ethnic minority affairs concurred, noting "Unfortunately, with a small axe, the wood cutter, year after year, has slashed and burned the forest for food production and finally changed the previously green map into a map with destroyed forests, eroded soils, dried out streams and rivers" (Cư Hòa Vẩn 1992, 1). A widely cited figure of 100,000 hectares of forest lost every year was attributed to swidden, without any firm basis of evidence to back this up (Hoàng Xuân Ty 1995).

Much can also be read from an examination of those who were not blamed for deforestation. SFEs themselves were usually exonerated, and the culpability of poor state management of nationalized forests was not recognized as a potential driver of forest change in any way. Officials often argued that SFEs were not the problem; for example, at a seminar on forest certification in 1998, a representative of the Forest Inventory and Planning Institute (FIPI) stated that "logging is necessary to supply wood to the national economy" and that "logging does not clear forest, but [does] create prerequisite conditions for [the] appearance of activities causing deforestation such as shifting cultivation [which] usually appears on the areas just after logging" (Department of Forest Development 1998). It was not uncommon to encounter reviews of forestry in Vietnam that simply fail to mention at all the existence of SFEs as forest managers over much of the remaining forest estate (VACNE 2004). Part of this invisibility may be the fact that the General Directorate of Forestry never produced a comprehensive map showing the location of all SFEs in Vietnam; I and other Vietnamese colleagues have asked repeatedly if such a map was ever made, and we have been told that while individual provinces have maps that show the locations of SFEs inside their borders, no national map was ever made to show the extensive scope of SFE operations. Just as certain maps ensured the visibility of specific types of forests and drivers of forest change, the absence of maps of SFEs has perhaps helped make these institutions less legible to criticism from donors or others.

By the late 1990s, newer drivers of and culprits for deforestation were emerging in the media, particularly cash crop development and spontaneous migration (De Koninck 1999). Coffee and other cash crops were increasingly linked to the felling of trees in the Central Highlands (Fortunel 2000), as high world prices for coffee starting in the 1990s combined with new land tenure rules that allowed individual farming by households formed a strong incentive for a land frenzy to expand cultivation.[6] Similarly, other cash crops, namely shrimp cultivation in the southern Mekong Delta, drove deforestation of coastal mangroves (Binh et al. 2005). But the culpability of the state and government policy in creating incentives for the expansion of shrimp and coffee farms, including some owned by state enterprises and local officials, rarely received as much attention (ESCAP 1991; ICARD 2002).

RESHAPING CONDUCT: RECLASSIFYING FORESTS

Given the myriad local and national understandings of and classification systems for forests, and the spiraling pressures on these resources, it is not surprising that a concerted effort to reclassify, visually map, and reorder management for Vietnam's remaining forests took place in the late 1980s.

This reclassification project coincided with the opening of relations with a number of former adversaries, culminating in the removal of a US trade embargo in 1995. The new openness extended not only to markets, but also to foreign relations, and international donors found willing partners in the Vietnamese state. Much of this turn to foreign aid from Western nations was in response to the need to find substitutes for the dwindling Soviet assistance of the late 1980s. The discovery of deforestation as a problem at the particular time in which donor aid became accessible therefore was not coincidental: defining a problem (deforestation) thereby created a particular solution (donor projects to help support an ailing sector).

Donor support for forestry went from very small amounts in 1986 to over one-third of the state forestry budget by 1990 (MOF 1991a).[7] Some of these projects focused on knowledge production for planning of forest development, such as a UN project that provided modern equipment for remote sensing and mapping of forest stocks to the Forest Inventory and Planning Institute. Other early donor supported projects included the development of a Tropical Forest Action Plan and support to design a Vietnam Conservation Strategy, both of which had a focus on sustainable forest management and which were modeled on larger strategies being developed at the global level (MOF 1991b). These and other projects encouraged the Ministry of Forestry to implement new systems of reclassification of forest lands that had first been proposed through Decision 1171 on the "Issuance of Regulations on Production, Protection and Special-Use Forests." This 1986 decree, coinciding exactly with the emergence of the *đổi mới* era, attempted to establish a classification system that would recognize more diverse interests in forests than state timber production, via land use planning of what activities could be done in what places and by whom. (This decision was later codified as national law in the Forest Resources Protection and Development Act of 1991.)[8]

The new classification system divided the country's forests into three specific types, based on management priorities: special-use forests (*rừng đặc dụng*), defensive/protection forests (*rừng phòng hộ*), and production forests (*rừng sản xuất*), ranging from more strict protection to less. This classification system was deliberately piecemeal, patchwork, and had something for everyone. New attention to forest conservation and biodiversity would be tackled in the special-use forests, SFEs would continue to have control over many production forests, and localities would be responsible for defensive/protection forests. The use of management categories, rather than ecological types, as the main organizing principle of forestry gives an indication as continued strong inclination to see forests as serving economic and ɔpment functions. Yet this new policy should not be seen as a retreat of the state from forest management; to the contrary, in fifty-four articles in the 1991 Forest Law, the "State" (*nhà nước*) is mentioned sixty-two times, and the state retained the primary role of defining forests, classifying them, and allocating management responsibilities to others.

Fully state-managed forest lands remained important, but in different ways than the timber-dominated SFEs of the past. The new category of protected special-use forests was considered a victory for a small group of scientists from the Comprehensive University (*Trường Đại học Tổng hợp Hà Nội*) and the State Committee on Science, who had long pushed for a category of protected areas that would focus on the preservation of flora and fauna in forest areas.[9] This group had managed to get one nature reserve, Cúc Phương National Park, declared in 1962.[10] Despite the scientists' lobbying efforts, it was not until after the end of the US war that a further round of ten protected areas, called at the time "prohibited forests" (*rừng cấm*), were dedicated in 1977.[11] The 1986 forest classification project expanded protected areas much more extensively, and established within the category of "special-use forest" the idea of national parks (*vườn quốc gia*), nature reserves (*khu dự trữ thiên nhiên*), and cultural, historical, environmental and landscape reserves (CHERs, *khu văn hoá, lịch sử môi trường và thắng cảnh*).[12] The total number of new special-use forests that were named in 1986 included seventy-three different local sites, with over half a million hectares proposed for protection (see figure 3.02). This rapid expansion of conservation zones in the 1980s mirrored other trends across Southeast Asia, given intensified interest in the region's forests from global NGOs concerned with wildlife and species preservation.

Yet there were other hidden reasons why the state expanded protected areas so rapidly in the 1980s and 1990, in a clear case of environmental rule. Most of the special-use forests demarcated in 1986 were formerly SFEs that were in financial crisis at this time, given declining state budgets under *đổi mới* (SRV 1999). With the new category of special-use forests, SFEs that were about to go belly up could be transformed into "national parks" and "nature reserves," and which were then eligible for international conservation dollars, rather than having to independently support themselves through timber sales in a new open economy. More than US$100 million in donor aid to conservation of wild animals, biodiversity and protected areas was pledged to Vietnam between 1986 and 2000, and much of this financial enthusiasm was driven by the discovery of several new mammal species previously unknown

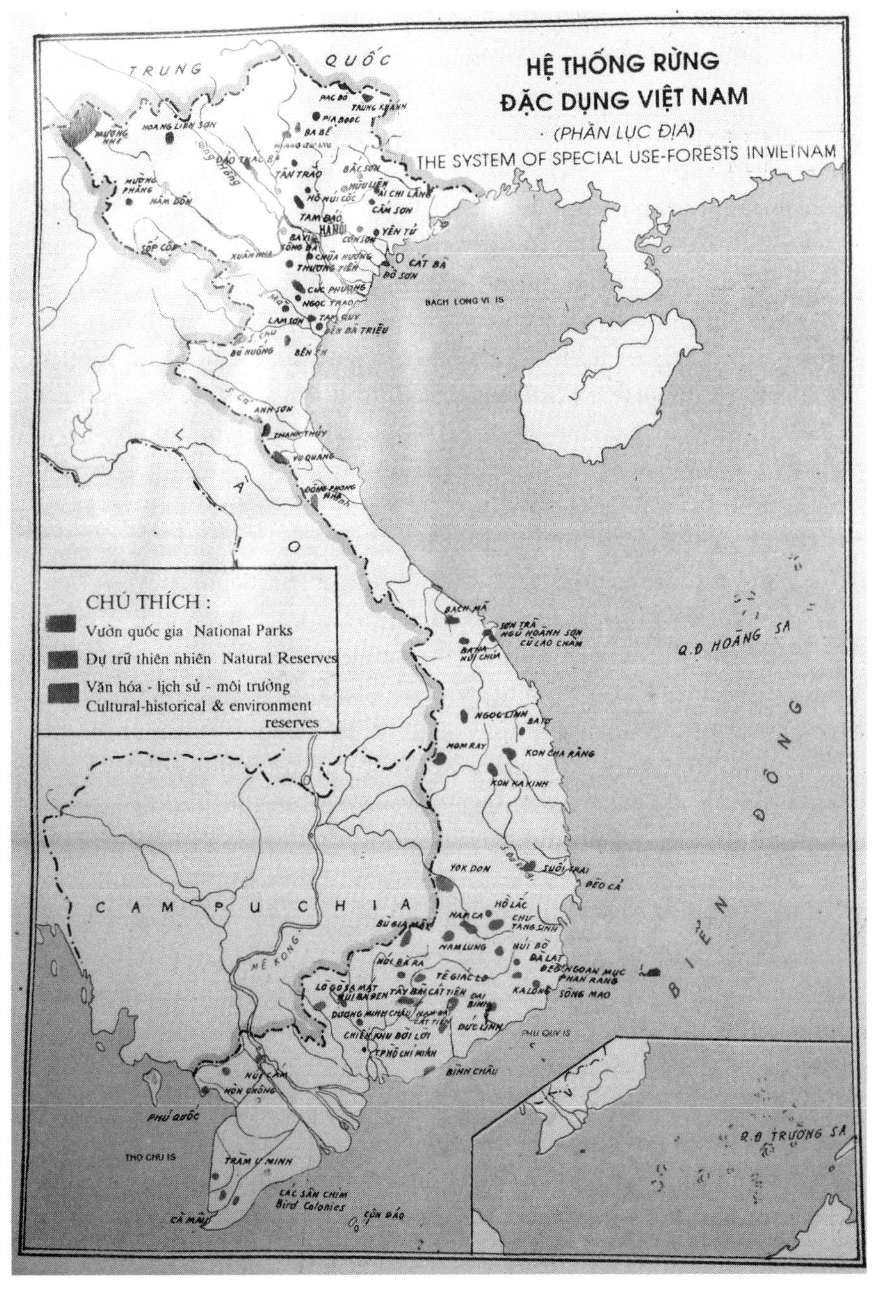

FIGURE 3.02 Special use forests in Vietnam, circa 1990. Map from MOF, 1995.

to science, such as the saola (McElwee 2002). This new interest thus created an incentive to turn SFEs into parks in order to access these funding sources. There was a further advantage from the former SFEs' point of view: many former SFE staff were also retained as state employees when parks were converted, allowing them to retain their pensions and benefits indefinitely (in other SFEs that had to close due to financial difficulties, SFE employees were laid off and often lost access to the guaranteed state pensions they had been promised). Thus despite the proclamation that biodiversity conservation was a new priority for Vietnam, in reality the location of protected areas was based largely on former SFEs, not on biological inventories or species richness. Nearly all major National Parks in Vietnam today are former SFEs, including Bến En National Park, Cát Bà National Park, Yok Đôn National Park (previously part of Ea Súp SFE), and Vũ Quang National Park (which became famous for being the site of new species discoveries of mammals in the early 1990s). Other parks may not have been logged by an SFE, but were the site of state agricultural farms, migration resettlement programs, or New Economic Zones, such as Bạch Mã National Park.[13]

The other new category of "defensive/protection" forests (*rừng phòng hộ*) were considered less environmentally valuable than the special-use areas, and limited exploitation activities could take place in these areas under the 1991 Law. All protection forests were to be locally classified and managed by Forest Management Boards run by state officials at provincial and district levels, while some land in these areas could be allocated to local households to manage under protection and reforestation contracts. Most of the area proposed for protection, around 4 million hectares, were watersheds, defined as areas with steep slopes over 25 degrees and a surface soil layer greater than 35 cm, with another 1.7 million hectares designated to support existing and proposed hydropower plants. A few thousand additional hectares were also designated along coastal belts and sand dunes (Hà Chu Chữ 1995) (see figure 3.03). As these designations indicate, the primary reason for classification of these forests was economic: to preserve downstream water supplies for agriculture and hydropower, or coastal areas for settlement and farming. Like special-use forests, most of the new defensive forests were located in dissolving SFE areas, with former SFE staff transformed into employees of "management boards."

The remaining forests (over 5 million ha) were to be available for production, including some natural forest areas, all plantation lands, and degraded lands to be reforested in the future. The production forests were to be managed by SFEs when necessary, but also potentially to households

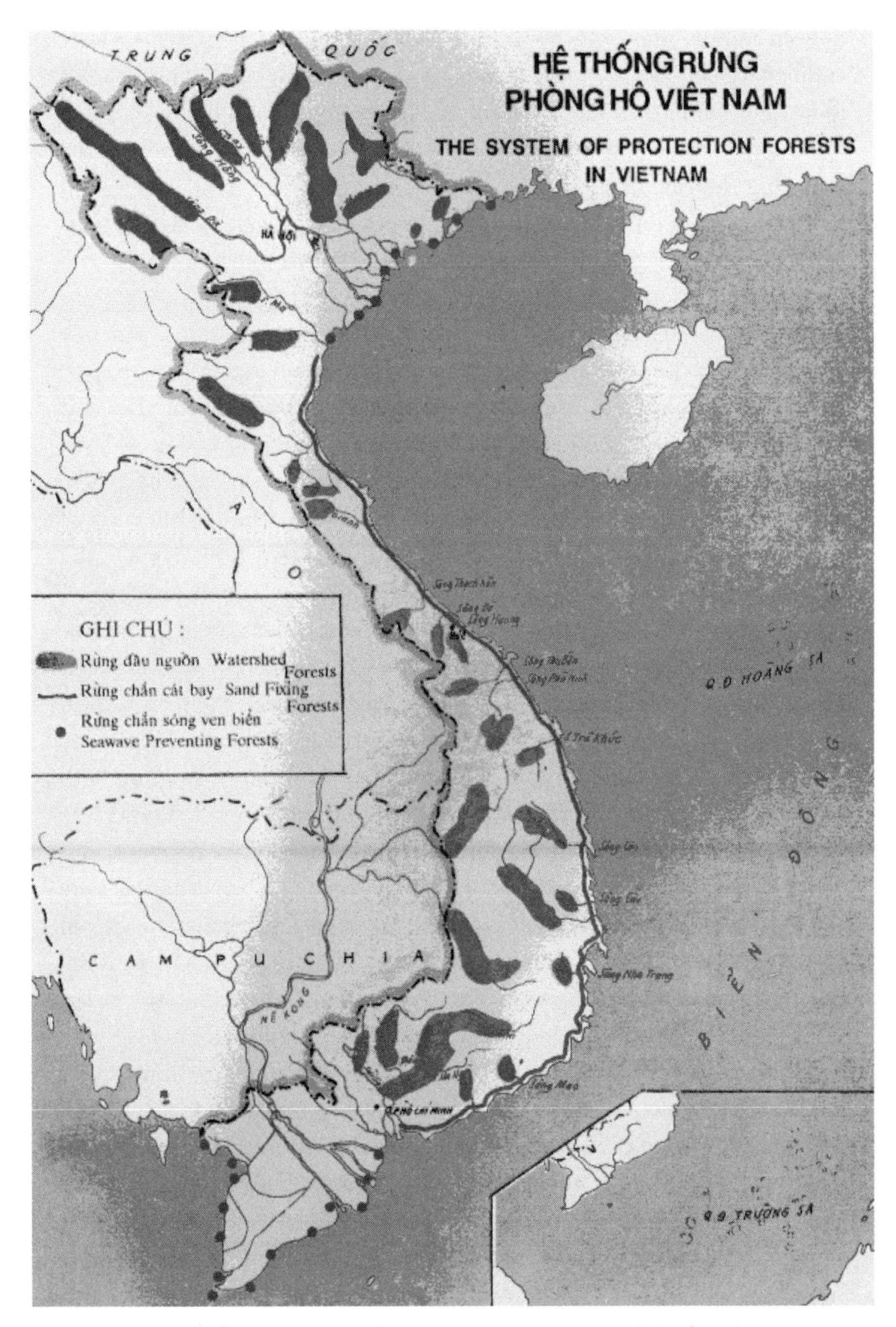

FIGURE 3.03 Defensive/protective forests in Vietnam, circa 1990. Map from MOF, 1995.

or even communities, which was the first time non-state forest owners were codified legally. The 1993 Land Law also confirmed the right of households to be involved in land management and led to households being allocated forest lands in the 1990s and 2000s, a process I describe more fully in the following chapter.

The three new forest categories were supposed to provide clear directions on where future logging could come from, while forests with biodiversity and watershed functions would be more strictly protected. But there was overlap between each category; areas were classified and then reclassified from one to another. The Kẻ Gỗ Nature Reserve provides a good example of this (see map 3.01). Parts of the KGNR had previously been in four different SFEs, and the area was logged heavily starting in the 1960s. Then the area was declared a defensive/protection forest in 1990, due to the presence of the Kẻ Gỗ Reservoir that provided water to the agricultural lands to the eastern coast of the province. In 1996, some of the protected forests were transferred to the special-use forest category and became the KGNR after lobbying by Birdlife International, which wanted the reserve to protect endangered pheasants. Of the 35,000 hectares within the KGNR, 24,801 hectares were classified as "special-use forest" and 10,500 hectares were still "defensive/protection forest." Yet local people often had no idea what category nearby local lands would fall into, and could not say if a neighboring forest was still an SFE, whether it had been turned into a new nature reserve, or whether it was something else entirely. Households could not sketch for me on maps who had authority over what lands and what activities might be permitted; they were told by all authorities that local use was illegal, particularly removal of logs. This left farmers on the ground very confused about where all the different land types began and ended.

To assist local people in understanding these new visions for forests, protected areas started putting up signs and boundary markers, reminiscent of the old French colonial forest service, who relied on handbills and tacked up laws to authorize rule. Often, boundary work at parks was paid for by international donor projects; in the KGNR, Birdlife funded new ranger stations along the edges of the reserve. Other protected areas engaged in different types of educational work to establish authority, focusing on creating new subjectivities for those who now surrounded the new parks. One common example was the distribution of "solidarity pledges" (*giấy cam kết*) to households, in which locals would be asked to agree to various rules. For example, around Đakkrông Nature Reserve, households were asked to pledge not to exploit, sell, or transport forest products illegally; not to hunt or trap wild

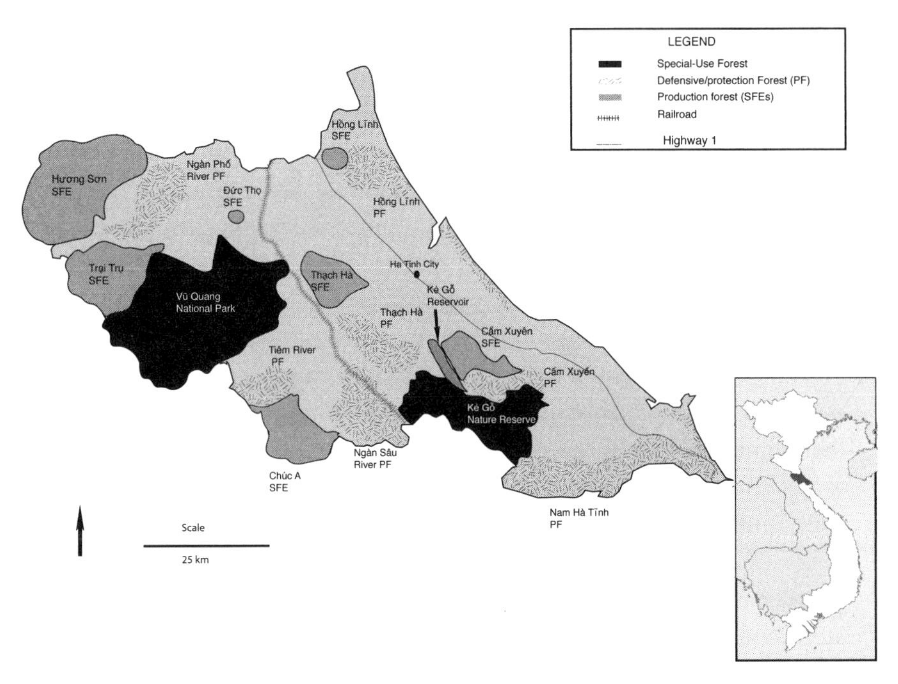

MAP 3.01 Reclassification of forest areas of Hà Tĩnh into special-use, defensive/protection and production areas, circa 2010. Map by author.

animals; not to protect others who did these activities; not to make swidden fields or forest fires; not to bring harmful substances into forest areas; not to fish in forest streams; not to graze livestock in forests; and not to use the forest to dig for minerals or unexploded ordinance to sell. These pledges were prominently displayed in local houses and public areas, and indicated that coercion alone would not be used to try to change conduct in protected zones in the new *đổi mới* era.

SHIFTING THE SHIFTING CULTIVATORS

There was also hope that the new forest classifications would assist in finally eliminating the practice of swidden cultivation. As the previous chapters have noted, various policies had been tried, all without success, to restrict this agricultural practice, yet these policies persisted. During the *đổi mới* period, the Fixed Cultivation and Sedentarization Program (FCSP) that had first been founded in the late 1960s underwent some changes; by 1995 the program was integrated with the New Economic Zones program under the Ministry of Agriculture and Rural Development (MARD). According to

ARD, there were three main objectives of the sedentarization program in the 1990s: 1) To reduce shifting cultivation and shifting residences and "settle" ethnic minority households; 2) to contribute to poverty reduction; and 3) to contribute to a reduction in deforestation and other environmental impacts.[14] The latter two justifications were new components that were added during *đổi mới*, and they were particularly aimed at enrolling foreign donors to provide money for FCSP projects. Targets of FCSP in the 1990s included 15,000 households that were said to still use extensive shifting cultivation; 4,000 households who were living in areas prone to disasters; and 20,000 households that were now living in special-use or protected watershed forests thanks to forest reclassification, and who needed to be resettled out of these zones (MARD 2004). For these latter households, they were targeted for movement (*di giãn dân*) to try to entice people to move away from protected forests and closer into villages. Yet the laws on forest reclassification did not include any overarching principles or mechanisms for dealing with resettlement for conservation, and these issues were generally addressed in an ad-hoc manner.

As a result of this confusion, I was asked by the Department for International Development (DFID) of the British Embassy in 2005 to assess the FCSP program to see if providing foreign aid support to these projects would constitute a violation of the often strict internal rules on resettlement that guided European donors. This work took my research team to two provinces, Gia Lai, in the Central Highlands where we worked with Bahnar communities, and Yên Bái, in the Northern Mountains where we worked with Hmong communities; in each site, we carried out surveys in both sedentarized villages targeted by FCSP and nearby non-sedentarized ones. There were no farmers practicing shifting residences in either area that we assessed, but that had not stopped these villages from being targets of the FCSP program. Rather, district and provincial authorities evaluated local households as being "weakly settled" (*chưa ổn định*), as they did not obtain high levels of production and were therefore considered at risk of returning to extensive shifting cultivation or migrating elsewhere. On the contrary, our research found that the targeted villages in Gia Lai in particular had been founded for at least a generation, and one informant reported that his first ancestors had arrived four generations previously. Yet in local government opinion, these areas were still not considered "strongly settled" (*đã ổn định*). In Yên Bái, there were no clear criteria used for targeting FCSP to particular households, other than the fact that the Hmong were assumed to be the best targets, as they were perceived as less integrated into mainstream Vietnamese society.

FIGURE 3.04 Agricultural-forestry mosaic of a Thái ethnic minority village in Yên Bái province, 2006. Terraced rice fields are near the valley bottom, swidden fields in the mid-slope (both fallow and active), and forest at the top slopes. Photo by author.

Areas of surrounding forest in both study sites were no longer freely accessible to local populations, primarily because of SFEs set up in the 1970s. In both sites, agricultural production usually included rotating plots of land in a fixed succession or rotating types of crops in cycles within a single field, neither of which contributed to deforestation as these practices were not extended into forested land but rather took place on long-used mixed-forest/agriculture plots (figure 3.04). With plot rotation, land was cleared and used until crop productivities began to decrease (usually in three to five years). At that point, the owner then left it for other lands, and after five to seven years, when the original land restored its fertility, the owner would come back for a new cultivation cycle. In no village that we assessed were households moving residences nor were they claiming new lands from forested areas or from other villages, and no deforestation could be attributed to their upland agricultural practices.[15]

Further, in our analysis of funding from FCSP, we were able to discover that zero dollars had been spent in either site on environmental projects; the FCSP monies had been spent for infrastructure development, construction of new commune centers, cadre training, residential resettlement, and production inputs, with no investment in any forest or environmental activities (Lê Ngọc Thắng et al. 2005). So if deforestation was not being stopped by the

FCSP, and the FCSP was not funding environmental activities, what was it supposed to be doing? In true environmental rule fashion, it appeared that FCSP was primarily about transforming ethnic minorities' culture, rather than nature. In Yên Bái, Hmong households that agreed to move down from a village at higher elevation were provided with 800,000 VND (US$60) to build a new house in the selected village site at lower elevation located next to the main district road, which put these households nearer Kinh authorities. In Gia Lai, FCSP money was used for re-planning of neighborhoods within existing village territory. For example, in one locale, forty-four Bahnar households were moved from their old village to a new one only one kilometer away, where they were given residential and garden land, brick houses, and some irrigated cultivation land, all of which resembled Kinh village habits and customs. The Bahnar villagers were forced to give up wooden stilt longhouses, which held multiple families (the traditional style), and to move into Kinh-style individual houses built on the ground, and made of brick with corrugated iron roofs. Such funding priorities seem to indicate that despite the reconceptualization to donors of the FCSP as a program for environmental purposes to protect watershed forests, much of the actual funding was spent on social goals of turning minority households into easily monitored village sites following a culturally Kinh model, just like earlier iterations of the program.

RESHAPING TIMBER ECONOMIES

In addition to reclassifying forests, setting up protected areas, and sedentarizing minority villages, another major policy action to combat deforestation in the early 1990s was an export log ban and new internal restrictions on logging. These actions were concurrent with other export log bans implemented in neighboring Asian countries, such as the Philippines and Thailand, at this time (Durst et al. 2001). The export ban was said to have come about as a result of then-Prime Minister Võ Văn Kiệt's trip to the central coast of southern Vietnam in 1990, where a number of wood processing enterprises were located. Quy Nhơn, one of the cities he visited, was estimated to have half the city's population relying for a living on timber processing, most of it illegally obtained, either by SFEs cutting above targets or by illegal loggers (AwBeng 1997). During his trip, Kiệt, who had a reputation as a more liberal minister than his predecessors, appeared genuinely concerned that wood processors were exceeding their allocated buying quotas. His administration quickly banned exports of raw logs in 1990, and raw cut and sawn wood in

1992 (Lang 2001). In fact, however, it was the domestic demand for timber that was much more important than exports in terms of driving deforestation. In 1989 it was estimated that 98 percent of the logs, 92 percent of the timber, and 89 percent of the paper that was produced in Vietnam was used domestically (MOF 1991a). Thus a logging export ban did little to curb this domestic need.

What did make a much more significant impact was a new Directive 90/CT from the prime minister, which issued a blistering indictment of "lax" forest management and instructed the Ministry of Forestry to "close" all special-use forests and defensive protection forests to any logging whatsoever. The logging ban in special-use forests would be permanent, and a thirty-year logging ban would remain on trees in defensive/protection forest areas. Commercial logging would also be prohibited in the remaining "natural" (*rừng tự nhiên*) forests in all of northern Vietnam and in the southeast and the Mekong delta areas, which were the areas with the least amount of forest cover remaining.[16] With 58 percent of natural forests now closed, only nineteen provinces would continue to have permission to do commercial logging, and only 105 SFEs or other entities would have logging permits, while the other 300 SFEs would be reorganized into protected forests or closed down. Future wood production was to come primarily from new plantation forests that would be developed (Durst et al. 2001). Each province was to come up with their own plan for "closing the forest gate."

Not surprisingly, major declines in forest profits occurred, and the total contribution of the forestry sector to gross domestic product, estimated at 3.0 percent in 1990, declined to 1.4 percent percent by 1995 after the bans (MARD 2001a). Timber production from the state sector (primarily from SFEs) dropped from 1,100,000 m^3 in 1990 to just 300,000 m^3 in 1998. The hope was that this lower level of natural forest exploitation could be sustained, while plantations and reforested areas began to grow so as to meet projected wood demand of nearly 10 million m^3 by 2005 (Durst et al. 2001). Yet the rapid reduction in just ten years from 12 million hectares of exploitable forests to only 5 million, while plantations had not yet developed fast enough to replace timber supplies, meant that the new restrictions on logging led to both a dramatic increase in imports from neighboring countries, particularly Laos and Cambodia, as well as strong increases in "illegal" timber extraction. In fact, just as the new rules on logging were being announced, a huge jump in timber exploitation occurred in 1993 as SFEs and other organization and individuals rushed to cut down trees before the new restrictions came into place. Further, an enormous trade in illegal timber imports rumbled to life in the late 1990s; one estimate proclaimed that the illegal trade in Cambodian logs to Vietnam was worth at least US$130 million per year (Global Witness 1999; 2000). Recent analytical work has relied on remotely sensed data and log export data to draw conclusions that Vietnam's logging bans and forest restrictions in the 1990s resulted in the displacement of a large amount of forest extraction to other neighboring countries (Meyfroidt and Lambin 2009b).

MAKING NEW SUBJECTS: ILLEGAL LOGGERS AND FOREST RANGERS

Despite the new forest classifications, the new protected areas, the new logging bans, and the new foreign aid, the mid-to-late 1990s actually saw an upturn in pressures on forests. Indeed, there is strong evidence that the interventions undertaken by the state to decrease deforestation actually had the opposite effect. The new forest classifications and closing of some production forests did not fall equally everywhere; forest restrictions were particularly hard on some provinces, such as those in the north, that saw all of their production forests close, but the new regulations were not accompanied by significant financial support to better patrol and protect these woods. The situation was further exacerbated by the rising demand for timber for a rapidly growing middle class who were benefiting from the *đổi mới* economic reforms. This timber could not yet be supplied by plantations, which meant that the incentives to ignore the new directives on restrictive logging were very high. The inaccessibility and rarity of certain woods, like *lim* (*Erythrophleum fordii*) and *pơ mu* (*Fokenia hodginsii*), also led to increased demand for these timbers as a status marker of wealthy urbanites (Hoàng Cẩm 2007).

The closing of SFEs without a clear alternative of where timber supply would come from, and the establishment of protected areas where no human use was to be allowed, left many rural households and enterprises in a quandary. How would they obtain timber? How could they survive alongside restricted land areas with less access to forest goods? Pressure did not cease on the newly protected areas—many of which were protected on paper only—and the SFEs that were dissolved reduced the national production of legal timber. Thus a gap formed whereby illegal loggers stepped in to fill the timber supply. Some estimates put the total amount of illegal logging occurring in the mid-1990s at 1 million m3 a year (Ogle et al. 1998). The enormous profits to be gained in the wood trade led to vulnerability of the system at all levels to corruption, tax evasion, and systematic cover-ups of deforestation.

The everyday enforcement of the new laws that were supposed tc deforestation fell to the Department of Forest Protection (*Kiểm lâı* theory, KL was a branch of the central government with a unified system of authority over localities, and was supposed to report upwards to the Ministry of Agriculture and Rural Development (also known as MARD, into which the Ministry of Forestry had been absorbed in 1994), but on an everyday basis KL officers actually were organized under provincial and district People's Committees and reported to local authorities. Hanoi was very far away from some provinces, as I was reminded by several front-line rangers. KL was supposed to enforce state laws on forestry, particularly regarding cutting and trade; be responsible for the oversight of many protected and special use forests; engage in management audits of SFEs and their approved cutting plans for production forests; and promote public awareness of forest protection.[17] In reality, KL officials found themselves squeezed by various constraints in carrying out these duties. They had very few resources for such a huge forest sector; they often clashed with local officials and resource users over who had authority over forests, some rising to the level of violent conflict; and most significantly, a series of high profile cases of KL rangers actually being responsible for illegal logging themselves came to light, which further undermined their credibility as a technology of environmental rule.

The first major challenge was the lack of resources for a professionalized forest ranger service. In 1979, there were 9,700 KL officers. In 1994, there were even fewer: only 8,400 rangers responsible for over 9 million hectares of forest. The head of KL in the 1990s, Nguyễn Bá Thụ, noted "Our force is very thin, missing people, and weak and our equipment is poor" (Cục Kiểm lâm 1998, 2). The lack of personnel for Thụ's department exacerbated the situation. It was not uncommon for a single ranger to be responsible for monitoring thousands of hectares of forest; in one example in Pù Mát National Park, the huge size of the park (nearly 200,000 ha) and small staff meant that each individual ranger was responsible for nearly 5,000 ha. The situation was slightly better at the KGNR, which had sixty-eight staff in the 1990s, forty of whom served as guards for 35,000 ha. But the sheer number of people who were now violating the new forest laws since the logging bans and reclassifications were instituted often overwhelmed the ranger forces. From 1995 to 2000, there were over 179,280 cases of "forest destruction, illegal exploitation, trade, transport and processing of forest products" that were documented, and undoubtedly hundreds of thousands more that went unnoticed, as any removal of forest trees without state permission was officially deemed to be "illegal" (Vietnam News Service 2000).

Yet while much of Thụ's tenure coincided with a crackdown on individual illegal loggers, he noted to me in an interview after his retirement that the larger problems facing KL were not these individuals, but structural problems: local provinces wanting to continue to use forest resources for local development, or to prioritize infrastructure like roads or hydropower over protected areas. The fact that new protected areas had once been used for exploitation meant that many localities wanted to continue to do so. Thụ listed several battles over protected areas that he had mostly lost: against the Daewoo hotel in Hanoi taking land from the national zoo; the Ho Chi Minh highway that was routed through Cúc Phương National Park; and plans to build karaoke parlors, a casino, and a train to the top of Tam Đảo National Park. Thụ emphasized that many of the development plans for conservation areas were put forward as "ecotourism" or "environmental development," but were really about lining the pockets of local elites. As he explained, "You cannot destroy a park like that for a casino: you destroy the principle of conservation. It's like you say you are selling a goat, but in fact you sell a dog; it's just a shell game (*lừa đảo*). For the provinces, they heard about 'what is ecotourism,' and they tried to do a shell game [to pass their plans off as environmentally motivated]."[18]

Another reoccurring problem was recruiting the right people to the KL and other forest protection services. For many years, KL rangers were primarily trained at the forestry universities in logging, as there was no educational program for conservation activities. Additionally, many former SFE workers overnight became forest protectors when watershed protection or special-use forests were designated, and found themselves lacking understanding of the tools that would be needed to orient their work to conservation and away from their experiences in logging. As the director of the Vũ Quang National Park, converted from an SFE in the 1990s, told me in an interview, "It is difficult to train people who used to be the exploiters to now be the protectors."[19] Further, most KL officers were sent to locations far from their home provinces, so they would not be tempted to help out relatives and friends in procuring illegal wood. However, this policy also ended up backfiring as local people distrusted the KL and saw them as outsiders; informants around the KGNR complained bitterly about the KL officers there, noting that hardly anyone even knew their names, as they had so little interaction with local villages. In interviews regarding interactions between locals and the KL, local farmers never used formal names to identify guards they had met, referring to the rangers as "Mr. Kiểm lâm" (*anh Kiểm lâm*) when they were being polite, and "Kiểm lâm bums" (*kẻ Kiểm lâm*) when they were not.

In addition to personnel challenges, there were also significant questions about the rights that KL officers held in order to stop illegal logging and the criminal and administrative measures that could be used. Of the 180,000 cases of forest crimes in the late 1990s noted above, only around a thousand ever went to a criminal trial. Instead, most crimes resulted in administrative fines or confiscation of the illegal timber, but these tended to be very weak enforcement mechanisms. Despite numerous policy orders from the central government, local rangers felt they had, in reality, very little legal support. The increased incidence of violence directed again KL officers made them additionally worried about forest patrolling; it was not uncommon to hear of stories of groups of fifty or more people with knives, crowbars, and hand saws uncovered by one or two rangers, who were unable to take any action against such large crowds. From 1996–2000, twelve KL officers were killed and 490 injured in clashes with illegal loggers (Vietnam News Service 2000). These incidents of violence even hit home in my own fieldwork; while I was in Hà Tĩnh in fall of 2000, a police chief in a neighboring district and his young daughter were killed by a homemade bomb that had been rigged by a sawmill operator involved in illegal logging.

Despite the rise in threats of violence, rangers told me that they were hesitant to use their own weapons while working in forests and often let people off with warnings and no confrontation if possible. A well-known case in Quảng Bình province in 2000 was discussed widely in the press and among rangers nationwide; the case involved a KL officer who had fired warning shots that had accidentally killed an illegal logger. The officer was prosecuted for murder for this action, and forced to pay almost $1,000 US to the family of the deceased, as well as $4 US per month as long as the deceased's children were young, and $2 US per month to his elderly father for the rest of his life (Phạm Hưng 2000). I happened to be visiting Ba Bể National Park at the time the verdict in the case came down, and I was able to discuss it with local KL rangers. Mr. Vi found the precedent in the case worrisome: "How can people carry out their duties if they are afraid of going to jail? He didn't fire first, the other one did. And he was sent to jail for some time. How can anyone else in Vietnam want to carry out his duties in that circumstance? If I go to the forest and meet someone with a gun, how can I do anything if I am afraid of going to jail?" Another ranger, Mr. Xuân, referenced a story he had recently read: "In America, they have solved this problem very well. There, the police have the total support of the government. They know they have the legal support. And so they are not afraid to carry out their duties and protect the forests strictly." Mr. Vi added, "we need the government of Vietnam to also ort us like that, and make the laws strict and clear." The fact that local forest users saw KL rangers as the instrument of the state (*nhà nước*), while rangers themselves felt weak and unsupported by state legal systems, gives an indication of the shifting modes of authority and subjectivity that the *đổi mới* era engendered.

Initially, KL rangers found themselves the object of sympathy in newspaper reports on illegal logging, which portrayed the noble Forest Protection Officer defending his green turf from the onslaught of hordes of loggers, and this patriotic image was commonplace among KL officers' own self-representation. One KL ranger I interviewed at the KGNR noted that he saw his job as providing a check on the self-interested nature of people who lived near protected areas, stating that "people exploit [the KGNR] very erratically (*bừa bãi*). They don't pay attention to the idea of sustainability of the forest (*sự bền vững của rừng*); they only pay attention to the sustainability of themselves (*sự bền vững của bản thân*)." The monthly Forest Protection Department newsletter distributed at KL headquarters even regularly printed a "Love Song for the Forest Protection Officer" to inspire the rangers: "Crossing over how many streams, through how many deep jungles, his tired feet toiling over the mountains and blind from fog, I hope you remember him… Planting trees on the mountains, stopping the swidden farmers, so the forest is always green, so the streams run clearly . . . our soldierly Forest Protection officer!" (Cục Kiểm lâm 1998, 27).

Contrasted with the heroic "green" rangers was a new kind of subject as well. Increasingly media and state attention focused on the figure of the "illegal logger," using a new term that had been coined to describe this character: *lâm tặc*. The word is a compound of the root for forest (*lâm*) and the verb "to hijack" (*tặc*), and is not found in any documents prior to the 1990s. Before the idea of a *lâm tặc*, people who deforested were simply referred to as that: *người phá rừng* (lit: people cutting forests). But the new *lâm tặc* became a particularly reviled figure of the 1990s.[20] This novel term seemed to encapsulate the official narrative of deforestation, in that a *lâm tặc* was one who hijacked the resources of the state in contravention of the revamped forest laws. Every week in the 1990s in newspapers published in Hanoi, it seemed there was a story about the nefarious deeds of *lâm tặc*. In one case in Quảng Ninh province in northern Vietnam, *lâm tặc* were so sure they would not get caught that they had invested significantly in developing their own logging operations, such as building permanent roads with dynamite, and going through and marking clearly the trees they wanted in over 1,000 hectares of a watershed forest (Nguyễn Thị Thanh Hải 1999). *Lâm tặc* with connections and power had various strategies for getting through the albeit weak forest law enforcement systems (for more details, see McElwee 2005). Because transportation of illegally cut wood usually required some sort of vehicle, *lâm tặc* were not usually the poorest people, but those with social relations and connections with officialdom. In many areas, local villagers were hired to be *lâm tặc* by local "fat cats" (*tay bụ*, lit: "chubby hand") to do the hard work in the forest, while the *tay bụ* acted as middlemen arranging the transportation of the forest products to trading depots outside protected forest areas. The local *tay bụ* then often reported to top-level mafia-type figures (known as "*ông trùm*," or "big men"), who ran national smuggling networks.

Interviews with *lâm tặc* that I conducted in Hà Tĩnh in 2000 revealed that use of vehicles with government or military license plates was a common strategy. Paperwork could be forged, if payments were made to the right KL office, allowing these vehicles to pass unimpeded through checkpoints. One estimate considered that twenty-three different officials would need to be bribed or persuaded to let illegal timber move from a community one-hundred miles outside Hanoi to markets in the city, indicating a high degree of graft (Sikor and Tô Xuân Phúc 2011). Indeed, many times it was not just forged paperwork or license plates, but full collusion of military and government officials that facilitated illegal logging. The perception that KL was not only impotent to solve high level deforestation cases, but was actually working in concert on these operations, was widespread.[21] In many areas, locals referred to forest rangers not by their official title of *Kiểm lâm* (literally meaning "forest checking"), but rather with a play on words as *Kiểm mâm* (meaning "looking for a large tray of food"), indicating concerns about corruption (Hoàng Cầm 2007, 26).

Certainly there were an increasing number of well-known media cases by the late 1990s that pointed out official state corruption in the forest sector. One of the most well-publicized cases was the trial of officials for deforesting in the Tánh Linh SFE in Bình Thuận province. Thirty-six defendants were accused of cutting down 53,429 m³ of trees in an SFE that was being converted to a protected watershed forest from 1993 to 1995, immediately after the internal logging bans and new classification laws for forests. The accused had made off with logs with a value estimated at US$1.5 million, and twenty-nine of the indicted were district and provincial officials from the Department of Agriculture and Rural Development, *Kiểm lâm*, the SFE, and other state offices (Long 1999). The culprits might have gotten away with their actions had it not been for a retired local forest cadre who harangued national officials with over seventy petitions alleging graft for more than four years before his complaints reached the president (Vietnam News Service 1999). In the end, all those associated with the case received jail terms, and the provincial People's Committee chairman was removed from his position. Similarly, smaller scale but locally well-known cases of forest corruption during my fieldwork in Hà Tĩnh province fomented an air of distrust around the KGNR.[22]

The fact that forest laws were overlapping and confusing meant that most people did not know what forest was owned by whom and what kind of extraction was legal or not; personal relationships were required for navigation of the byzantine forest laws. And because many KL did not have kin or marriage relations with local people, having been stationed far from their home provinces, they formed economic ones. Everyone I spoke to believed that any KL guards could be bought off for the right price. KL rangers had the legal right to punish loggers in the KGNR area with confiscation of timber discovered and any equipment used to cut it, as well as a monetary fine up to 500,000 VND (US$30). But this could be avoided with the right payment. Most rangers earned very little on government salaries, around US$30–$50 per month at the time of my fieldwork, and so it was easy to bribe them to turn an eye to small-scale local extraction. A bribe of 300,000 VND (US$15) was the going rate to extract timber worth about 1 million VND (US$66) when I was doing fieldwork, and this act of giving bribes to local rangers was known as "making laws" (*làm luật*). One district agricultural officer in charge of reforestation told me that there was no point to my research on forest management at the KGNR because "there is no longer any forest. The people have deforested it all. Nothing left. Kẻ Gỗ is just planted acacia and pine, the rest of the forest is gone. And when you see who protects the forest, you aren't surprised. When anyone can 'make laws,' and get timber, you just have to pay an 'entrance fee' to go to KGNR for logging." After I left Kẻ Gỗ, the press exposed a major scandal involving KGNR rangers in 2007 in which the ease of bribing was described in detail, although it apparently led to little change (Anon. 2007).

BECOMING A *LÂM TẶC*

In addition to large-scale logging gangs, much of the so-called "illegal logging" in Vietnam was being carried out by millions of small-scale loggers. To learn more about the economics of forest use and logging, during the course of my fieldwork in the KGNR in 2000, I surveyed 104 households, a random sample of 20 percent of the residents of five main villages in the buffer zone, on their forest use through in-depth interviews. While some households

FIGURE 3.05 Small-scale illegal loggers (*lâm tặc*) leaving the Kẻ Gỗ Nature Reserve, 2001. Photo by author.

could meet their timber requirements through home gardens and private forests, most households had no such legal access to forest lands (75 percent of surveyed households had no private forest lands at the time), and were forced to rely on wood from either the nearby SFE or the KGNR. Yet continuing to use forests near their homes for wood, activities which had once been freely available and even encouraged in the 1960s and 1970s, became illegal with new forest classifications, and signs were put up around the perimeter of the KGNR warning people to stay out. As one nearby resident stated, "In the past we were free, but now we are prohibited from the forest" (*Trước là tự do, mới cấm rừng*). Another man I interviewed simply added, "It's the state's forest now, not for us any longer" (*Rừng quốc gia, không làm nữa.*)

In addition to subsistence needs for timber, 12 percent of the responding households to my survey were engaged in illegal logging activities for commercial purposes, and they received on average 15 percent of their income from timber sales. During field trips to KGNR, I myself discovered *lâm tặc* on two different occasions. Once, a group of eight men were coming out with four bicycles full of sawn timber and handsaws (figure 3.05). They had about twenty planks of fairly poor quality wood, which they estimated they would sell in the nearby district market for 50,000 VND (approximately $3.30 USD) each. At that price, each *lâm tặc* in the group would have earned no more than seventy-five cents US per day for this logging trip.[23]

Comparing the (thirteen) logging households in my sample versus the (ninety-one) non-logging households, there were no statistically significant income differences between the two groups. So if it was not necessarily poverty that pushed some households to engage in illegal logging, what was the driver? For some, the ease of access to the KGNR was a draw. As a male logger in this thirties said to me, "Whoever wants to cut trees, goes to cut trees. Whoever wants to deforest, goes and deforests" (*Ai chặt thì chặt, ai phá thì phá*). The activities of rangers made little difference to him, and households with excess male labor particularly found illegal logging a low-risk option for employment. Villagers who lived farther from ranger check points not surprisingly tended to have higher numbers of people engaging in illegal activities. A man at a group meeting in a village physically located near a key KL checkpoint summed up the dilemma: "People in this village have two problems; one is that we're close to the authorities and two is that we're far from the forest." He went on to explain how other villages far from KL stations took advantage of the access: "How many people go to make charcoal to build nice concrete houses!" Another headman in charge of a village nearest the KL checkpoints told me,

> From long ago, this village was near the forest (*sát rừng*). People here know the forest, since they grew up with it. Like me, ever since I was ten years old I went with my grandfather into the forest to go get wood, or fuelwood, or to hunt. Over in the forest, I don't need a map at all. People from the lowlands [*he meant migrants to the area, some of whom had worked for the SFE*] don't know the forest as well as people here, because people here have lived with the forest their whole life. Before the Kẻ Gỗ Lake was built in 1976, the forest was right here [he pointed outside.] You didn't need but to step outside your house, and to get valuable wood to build a whole house. It only took me a day to find wood and cut it for mine.

Now, he said, other villagers were cutting all the timber down and exploiting the KGNR irrationally. When I asked why people in his village didn't just assert claims on the forests, or take their case to a higher authority, he said sadly, "Because we don't know the landscapes there now, and it is far away. The forest has really receded. Also it is because we have to pass the commune people's committee building to get back home from the forest. Others from other communes go into the forest, but they don't have to pass the authorities to do so."

Others who were now "illegally" using local resources made a moral claim on the forests of the KGNR and argued for notions of social justice in local rights to trees. They would point out that as farmers, they were relatively poorer than KL guards, who at least had regular government salaries and pensions, and that the so-called "illegal actions" they were accused of could be justified. Subsistence claims—such as the need for timber for personal house building—were regularly made by farmers to justify their so-called "illegal" log taking, and they vociferously pointed out the non-commercial nature of their activities to me. The fact that farmers regularly saw others with higher status removing logs, and that rumors circulated wildly about which officials were involved in authorizing such activities, created a situation that was described to me by use of a common phrase in Vietnamese: "*trâu buộc ghét trâu ăn*" (the buffalo that is tied up hates the buffalo that is grazing).

This was not to say that no one around the KGNR was concerned about deforestation. Indeed, many households expressed worry about the rapid rate of environmental change outside their doorsteps. A man in his sixties noted that "We used to be able to find trees as big as I am; now the forest is badly degraded (*rừng bị hư*)." One middle-aged man said in an interview, "In the past, before when we didn't have a house yet, we went to the forest a lot. We weren't the only ones; everyone was cutting at rates that were just too high. Back then we still had lots of rarc and valuable trees, but now when you look at it, it is all bare lands and bare hills. Now if people keep exploiting the forest excessively like we used to do, in two years it will all be gone." Another fifty-year-old man added, "The forests are worse off now, thanks to people exploiting them. Now all we have are small young seedlings trying to grow up. There are only a few kinds of trees left to develop back into forest. We've destroyed the large trees completely already." A few interviewed households expressed muted support for the idea of a protected area, given this deforestation. When I asked a thirty-nine-year-old man living near the new KGNR border what he thought of the park, he gave a mixed answer: "I like that it is protecting plants, water sources, making things greener, regrowing forests, but on the other hand, I'm losing my income." Another neighbor pointed out that more support would be given by locals if the KGNR had some sort of way to create income to substitute for that lost to increased forest restrictions. "When people have no other work, they go into the forest. This is not a large income—we are not getting rich! We are just keeping our heads afloat." Another respondent, when asked if he liked the KGNR, said that although it was "difficult for me personally, I see that it has benefits for the country."

For many of those who were concerned about deforestation, they felt helpless to do anything about it; these citizens were informed they needed to personally conduct their conduct differently, according to the new forest laws, but they were not included in any type of local co-management or consultation with the KGNR staff and rangers. With aid from international donors, such as Birdlife International, the KGNR had started educational campaigns to alert locals to the presence of the new boundaries and forest classifications, using radio announcements and posters throughout the villages on the border of the reserve. The message of these campaigns was that the KGNR was state forest property and any entry into the reserve would be considered illegal trespassing. While these campaigns attempted to instill new subjectivities in nearby villagers, what they also did was reinforce the idea that only KL rangers had the right to determine who could use the KGNR and who could not. One village headman next to the KGNR entrance noted to me that the most commonly used path by *lâm tặc* from a differing district to access the KGNR ran right through their village agricultural lands. He stated that it would be easy for his village to stop and catch these loggers, but that the state had made it clear that the KGNR was the "government's property" and business, not theirs, so they made no attempts to do so. One elderly man living near the KGNR summed up the management problems for me with a well-known phrase in Vietnamese "no one cries for the father of everyone" (*cha chung không ai khóc*); in his opinion, the state ownership of the KGNR had removed any incentive for locals to get involved in co-management, and if the KL didn't seem to care much that deforestation was happening under their noses, why should the nearby farmers?

While the citizens around the KGNR largely protested their exclusion from forest resources through ignoring forest boundaries and continuing to engage in small-scale forest extraction during my fieldwork in 2000–2001, a few years later, some households actually took the step of occupying nearly 100 hectares of the KGNR in protest against lack of access. These types of protests have taken off in other areas of the province as well, particularly over forest lands owned or managed by the state, including 150 hectares disputed between the Hương Khê SFE and nearby households, and 350 hectares disputed between the Ngàn Sâu defensive/protection forest management board and a local youth pioneer group that believed it had been allocated the land to protect and manage (Hùng Võ 2013). These protests were an indication that the individual subjectivities that were supposed to be fostered in the *đổi mới* era by government campaigns on conservation and new protected areas boundaries could not overcome senses of injustice and inequality over land rights. More and more households decided for themselves that they should have rights to forests, leading to both new forms of conflict and a rapid increase in individual ownership of forest lands, a topic taken up in the next chapter.

CONCLUSION

Market reforms and trade liberalization have opened new opportunities in the Vietnamese economy, resulting in dramatic transformations in social and economic life. But it would be simplistic to blame market forces alone for the deforestation that occurred in the 1980s and 1990s as Vietnam opened up to the West after years of isolation. While some clear drivers of land use change can be linked to the opening to the global economy, such as coffee expansion, other problems, such as conflicts between local populations and SFEs and other state management agencies, were long-standing.

Yet the policy responses to deforestation often focused on single drivers, such as export of raw logs, or shifting cultivation, without acknowledging the multiplicity of causation and the temporal dimensions of changes. Too often, policies enacted during the *đổi mới* period that attempted to control deforestation were premised mostly on proximate actors—such as illegal loggers, agricultural migrants, and shifting cultivators—rather than the underlying causes, which included local distrust of state forest institutions and competition over land rights, national policies which encouraged migration and cash crop development at the expense of forests, "shell games" over development in ostensibly conservation-focused areas, and the continued control by SFEs of too much land area with insufficient management, leading to overcutting and poor monitoring and protection.

These complications meant that environmental rule during this era was of a different kind and scope than had been seen under the French and the DRV. Government campaigns to "educate" local populations regarding the new zoning and the state's role in protected area management, as was the case around the KGNR, demonstrates the shift from disciplinary forms of forest management that had previously been embodied in SFEs, to forms of rule that depended more on individual subjectivities, such as the need for conservation education and pledges to undertake biodiversity protection. In the gaze of the state, the revamped forest classifications were a clear path to these new ways of conducting conduct, but in reality, locals subjected to changed classifications did little to alter their attitudes or practices. Indeed, the central state efforts to control forests and logging through forest reclassification, state management, and logging bans created a vacuum that was filled by corrupt officials and wholesalers who ended up being the prime beneficiaries of these policies.[24]

Policy responses to perceived deforestation in turn created feedbacks and drivers of still more land use change. The constricted supply of logs that was a result of logging bans, combined with the new open market forces of *đổi mới*, created an extraordinary demand for high quality timber, and this in turn created a boom in incentives for illegal timber production. The reclassifications of some forests into protected areas and restrictions on local use and residence in these zones also resulted in local conflict. The spatial reclassification of forests led to the simultaneous social reclassifications of forest-using peoples into categories of "illegal loggers" and "forest destroyers," even when twenty years previously these subjects might have been lauded for helping to convert forests to farms during socialist frontier expansion.

The technology of rule that predominated in the *đổi mới* era's approach to forests was the individual forest ranger as the material representation of state authority. These agents served as physical objects, which could be cursed or attacked or obeyed, as the case might be, as well as new types of embodied performers, who could be ascribed different types of subjectivities and motivations. In rangers' own eyes, they were the only force that stood between state forests and greedy local populations, but who had little authority and support from the larger state. In locals' eyes, the rangers were individuals who could be corrupted, as well as working for a system that was untrustworthy and arbitrary. The conflicting positionality occupied by the rangers points out the need to pay attention to networks of social relations in understanding environmental rule, rather than simply looking at laws and policies implemented from the state.

Average citizens found it hard to stomach that forest access was restricted for the poorest, while corruption and a lack of genuine concern about deforestation from state agencies seemed the norm. If the government was so interested in forest conservation, why did it do such a poor job of guarding it, wondered many locals. This skepticism about environmental rule ultimately proved to be its undoing. The perception that the state, in the form of individual KL rangers, was both the oppressor of local resource rights and the cause of illegal deforestation prevented the grand new plans for forest protection and protected areas management from being effective across much of the country, leading the deforestation crisis to get progressively worse, not better, throughout the 2000s. Even later forest policy approaches that tried to recalibrate state-society relations, namely land tenure allocation and reforestation projects (chapter 4) and payments for environmental services approaches (chapter 5) to individualize conservation and take it out of the hands of the state, have been unable to overcome many of the challenges of the *đổi mới* era that hardened perceptions that the government role in forestry was often incompetent, illegitimate, or unnecessary, and should be contested. This loss of moral authority by the state meant that no piece of paper declaring forest laws, or boundary on a map showing protected areas, was likely to be effective in changing forest use by citizens, who refused the roles of complacent environmental subjects in favor of their own definitions of rights and authority over forests.

4 Rule by Reforestation

Classifying Bare Hills and Claiming Forest Transitions

WHEN I FIRST BEGAN TO DO FIELDWORK OUTSIDE HANOI IN 1997, after language study that left me with passable but limited Vietnamese, I often found myself warmly welcomed in rural areas once I had announced my name. It took a little while to understand villagers were associating me with a large donor project for food, cash, and tree seedlings in rural areas in the 1980s: the PAM project, named after the World Food Program's French acronym. Many households in midland areas outside Hanoi and along the northern central coast were paid 500kg of rice in return for planting a hectare of forest, and these households remembered the food support as having been crucial at a difficult period. They were always eager to thank me for "my" project that had provided them with an extra bowl of rice at mealtimes. However, when I pressed these households about the details of the tree-planting projects, they had less information and understanding about what the goals and objectives had actually been. That these households remembered rice support and food security, and not the actual seedlings they were supposed to have planted or what they had done with these trees, provides the backdrop for this chapter's exploration of tree-planting initiatives as a form of environmental rule, which have often been more about land, social relations, food security, employment, and other factors than about the expansion of tree cover itself.

In addition to the PAM project, other national initiatives since the 1950s have tried to increase Vietnam's percentage of total forest cover, the most well known being the Five Million Hectare Reforestation Project (5MHRP) that ran from 1998 to 2010. These initiatives have usually combined extensive state investment in seedling provision with some form of land allocation (either temporary use contracts or permanent privatization) to households to encourage them to plant trees. Such programs often focused on afforesting lands the state problematized as "barren," and recipient households transformed these lands into smallholder forestry plantations, ostensibly to reap both environmental and economic benefits. Vietnam increased from less than one million hectares of plantation forest in 1990 to 3.5 million hectares by 2010, an afforestation rate of 150,000 hectares a year, the third highest globally in the past decade (FAO 2010).

Subsequently, Vietnam has been held up as a country that has made a "forest transition," moving from net deforestation rates to overall net afforestation rates. However, this so-called forest transition has been controversial; recent remote sensing work raises doubts if Vietnam ever made a transition at all, and even politicians within the country have questioned the official numbers about forest cover gain. Assessing claims over forest transitions as a project of knowledge construction to further environmental rule helps us see why this process has been so controversial, and why there may never be a resolved "truth" about exactly how much forest cover Vietnam actually planted in the 1990s and 2000s.

Environmental rule also helps explain why the state created and supported such large-scale tree planting policies at this time. One major factor was a concern for food security, although not for the households asked to plant trees: rather, the concern was having upland forest cover to supply water to lowland agriculture, and this concern over rice production stemmed from continued fears over famine since the 1944–45 disaster, along with a legacy from the French era that upstream forest cover was the primary influence on downstream water supplies. A further strong impetus for forest plantations was the army and Party ownership of thousands of small-scale sawmills suffering since Prime Minister Kiệt's logging bans in the early 1990s, and which needed new supplies of timber.

These tree-planting projects particularly took off after the introduction of *đổi mới*, which ushered in a new era of relations between the state and the individual citizen-subject regarding land management. Whereas the 1990s had seen conflicts over protected areas, rangers, and access to trees, the 2000s saw a refocus on afforestation policy that was more about land rights than trees themselves. The reforestation campaigns reflected new concerns about state-society relations in a market economy, with households stepping up to become individual capitalist land managers again after years of collectivization and socialism. The tree planting campaigns were an ideal intervention to change conduct and to encourage responsible land ownership and use, as nature became a site of reenactments of nationhood and subject formation.

Yet despite the not-necessarily "green" reasons driving the reforestation programs, many environmental NGOs strongly supported these efforts. Environmental actors become engaged in the network to support afforestation of lands with exotic species, despite evidence suggesting more biodiverse natural regeneration was eliminated in the race to expand new plantations. These reforestation projects became a key moment of "translation," as international donors were enrolled in networks of support though the use of persuasive tropes and discourse about "regreening" barren lands, and through the use of visualizations, including a map purporting to show extensive forest cover in 1943 that had subsequently been lost, but could be regained. The government declared the end result of the twelve-year replantation campaign a success, with rosy statistics about the increased amount of land with tree cover; these sites were subsequently called "forests," although this was a slippery definition, given the mixed environmental benefits of afforestation with non-native trees.

Local households had their own reasons for planting trees, ones that had not been foreseen by authorities, and in this they ended up subverting environmental rule for their own purposes. Households readily planted tree species they had no particular economic nor environmental interest in—namely low-value eucalyptus and acacias—largely to claim land during a brief phase when the country was radically shifting from collective to privately held land ownership. For many locals, the tree-planting programs became an opportunity to exert land claims at a time in which there was a sense of openness and uncertainty about what areas belonged to whom. This allowed new claims to be registered through the act of planting trees without seeming as selfish as if the household had planted crops on these lands for households' own use. The trees, after all, were being promoted as a "national benefit."

These changes in land rights and claims were enabled by the problematization of vast areas of the rural countryside as "bare hills" (*đồi trọc*) and "wastelands" (*đất trống,* lit: empty land). Variations on the idea of barren land had long been used in Vietnam, dating back to the colonial era, to define vast stretches of property as places the state would like to see as forest; these areas were classified as degraded, valueless, ownerless, and in need of environmental rehabilitation. It is clear the problematization of lands as "bare hills" directly defined the technical intervention proposed: make these lands unbare through afforestation. In reality, these lands were already sites of natural regeneration and active use, often by those households without significant forest land holdings of their own. As bare hills were privatized and turned into smallholder plantations, the poor became the least likely to receive land allotments.

Trees were important actors in this story in their own right. The knowledge produced about bare hills by both national and local forest officials—that trees of any kind were better than no trees—led to incentives to plant cheap, fast growing species over more slowly growing indigenous ones. The two major genera chosen were exotics with high water demands and low quality timber, and the biological characteristics of these trees created specific types of policy demands and needs for changing conduct, as they did not regenerate readily. They had to be planted, over and over, and thus required greenhouses and technical expertise for growing the seedlings, and labor for reforestation. As a result, they drew humans into the practice of reforestation in specific and mutually constitutive ways.

GENEALOGIES OF REFORESTATION POLICY IN VIETNAM

Identifying "degraded" lands in Vietnam's countryside and reforesting them had long been a goal of successive administrations. French colonial foresters put in place the earliest reforestation programs, as noted in chapter 1, given their serious concern that overexploitation of the most accessible lowland forests without proper regeneration had led to the creation of deforested lands, resulting in soil erosion, landslides, and changes in water and climate. Part of the Cochinchinese Forest Law of 1912 addressed this problem, giving the Forest Service the right to require afforestation, even on private lands, in cases where "conservation is necessary"; such reforestation was deemed imperative in circumstances including:

> 1) In keeping land on the slopes; 2) In defense of land and banks against erosion and encroachments of rivers; 3) Along water sources or streams; 4) For protection against coastal dunes and the erosion of the sea and sands; 5) For defense of territory in the border area to be determined, if necessary, by order of the Governor General; and 6) Public safety. (Sarraut 1913)

By the 1920s, the Forest Service began to concentrate efforts on replanting areas consider to be most degraded, or having poor stocking densities of trees. Reforestation focused on central Vietnam near urban areas, such as Huế, Tourane, and Vinh/Bến Thủy, where it was becoming increasingly difficult to access firewood (Guibier 1918). An experimental station in Cochinchina focused on generating expertise in reforestation, and undertook trials in regeneration, focusing on *phi lao* (*Casuarina equisetifolia*) for coastal dunes and pine species for the highlands. However, the forest service faced a number of challenges, including the high expenses required to propagate seedlings and then pay rangers to replant them, as well as the resistance of tree species themselves to regeneration (Guibier 1924). Saplings did not like to grow in the dry hot Indochinese sun and needed to be under canopies, necessitating the use of "trash" but fast growing trees to shelter good hardwood species while growing (Guibier 1918). "The problem has many technical difficulties and great expense must be considered due to distance. We have not yet found many species likely to give cost-effective results in pure stands, especially in soil already depleted by more or less intense deforestation" admitted the Forest Service (Service scientifique de l'Agence économique de l'Indochine 1931, 8). The only species regenerating well in "pure stands" was *phi lao*, and consequently eucalyptus seedlings were imported from Australia for experimental purposes in the 1920s. Because the focus on state-led reforestation by "experts" and rangers was advancing so slowly, the revised Indochinese Forest Regime (*Régime Forestier*) of 1930 first permitted the acquisition of land by planting of trees to claim ownership (Title V, Articles 74–82). Yet despite this attempt to encourage smallholder reforestation, the French reforested no more than a few thousand hectares by 1940 (Gourou 1940).

Later reforestation efforts would be equally slow. Throughout the early years of the Democratic Republic of Vietnam when the SFEs were first established, they were supposed to be reforesting the lands they harvested, but rarely did so. For example, from 1961–65, the Hương Sơn SFE cut 200,000 m³ of timber, but only reforested five hundred hectares (Văn Ngọc Thành 2005). Though concern was expressed this style of management was akin to "eating seed corn" (*miệng ăn núi lở*) and that if the state did not actively replant forests they would gradually be destroyed, reforestation efforts at SFEs remained abysmal (Lê Trung Đình 1967, 61).

Partially in response to these failures, smallholders became targets of intervention as reforestation was encouraged outside of SFEs; for example, in honor of the thirtieth anniversary of the founding of the Communist Party, President Ho Chi Minh started a New Year's tree planting tradition (*Tết Trồng Cây*) still important to this day (figure 4.01). Ho presented the movement in an article in the November 11, 1959, issue of *The People* (*Nhân Dân*) newspaper, calling on each citizen to plant one new seedling and care for at least four trees total in the first month of the Lunar New Year each year. He argued this would lead to at least fifteen million trees being planted annually, which would provide useful wood, beautify the landscape, and regulate climate (Nguyễn Văn Công 2001). By 1961, these goals had been ambitiously expanded: each year, nearly 100 million trees needed to be planted, including

FIGURE 4.01 Ho Chi Minh encouraging tree planting. Photo from *Nhân Dân* [The People] newspaper, undated (1960s).

a precise number of 90,535,800 trees for protection and fuelwood purposes, 328,000 street trees in urban areas for beautification, and 7,495,000 trees in SFEs. In mountainous areas, the Lunar New Year's campaign was combined with movements to convert swidden fields into tree farms and nurseries for tree seedlings, and localities were to have contests over who could plant the most.[1] From 1960 to 1965, the Lunar New Year movement is estimated to have planted 600 million trees (Nguyễn Tạo 1968).

Government officials encouraged the expansion of these reforestation initiatives, around houses, along coastlines, in highland watershed areas—any place not directly supporting agriculture should be a tree farm. This was especially the case along rivers and coasts, as "planting trees to afforest dikes to protect rivers is a good method that is inexpensive, at the same time providing a source of wood, firewood, bamboo, bark, etc. . . . serving economic interest, improving lives of the people living along river banks and coasts."[2] The National Irrigation and Agriculture-Forestry Departments were told to work together to improve reforestation, explicitly linking the goals of forest cover to the protection of agricultural crops downstream from the twin threats of drought and flood.[3] In this belief, the DRV leaders took ideas about forests and hydrology straight from the French colonial forest service.

Later, however, the afforestation program was promoted for more militarily strategic reasons, as reforestation would provide hiding places or training spots in villages for citizens to fight the US enemies (Nguyễn Xuân Khẳng

1965). In 1969, just before his death, Ho Chi Minh wrote that tree planting was a patriotic act, as forests "bring large benefits to the economy and to the national defense." He added:

> In the years we have been fighting the destructive war against the Americans, our people have seen clearly the benefit of planting trees, so our tree planting customs are developing. The act of planting trees brings continued benefits like wood, fruit, and windbreaks so we can plant rice, protect dikes, cover hills, prevent soil erosion, prevent sand, etc. Those trees that we have planted along the sea, along roads, around our villages, all have large benefits.[4]

People who excelled at tree planting, like other patriotic activities, were to be given the designation "hero" (*anh hùng*). After Ho's death, the party decided that year's Lunar New Year planting would be renamed the "Tree Planting for the Lunar New Year to Resolve to Win over the American Invaders" (*Tết trồng cây quyết thắng giặc Mỹ xâm lược*), while the following year's movement would be named the "Tree Planting for the Lunar New Year to Always Remember Uncle Ho" (*Tết trồng cây đời đời nhớ ơn Bác Hồ*).[5]

These initiatives primarily relied on voluntarism to plant trees, with small amounts of funding per tree planted, and the good will of people to keep them there. Without additional incentives to retain forest cover, however, these reforestation efforts began to run into difficulties. The total number of hectares reforested in the North at the close of the Vietnam War was less than 100,000 hectares (Hoàng Hòe 1996), and the situation was even worse in the former South Vietnam, where, despite government investment in seedling greenhouses and funding to put refugees to work planting pines, teak and casuarinas, only 15,000 hectares of land was reforested in 1974 (Swanson 1975).

REFORESTATION IN THE POST-WAR ERA

Given the continuing slow pace, yet the increasing concern with degraded lands, especially the many areas of the former South Vietnam affected by war damage, reforestation became more urgent in the post-war era. The issue was important enough that the Fourth Party Congress, held in late 1976 immediately after reunification, declared the Communist Party would "attach much importance to all problems of reforestation, exploitation of forests, and to covering all the hills and dunes with trees during two or three cycles of Five Year Plans; to launching a movement of all the people for the reforestation and protection of forests" (Võ Quý 1985, 100). Experimental sites for reforestation were established in several forests, like Mã Đà in southern Đồng Nai province and the Cần Giờ mangrove reserve outside Saigon, both of which had been hard hit by wartime defoliants. The difficulties of establishing new trees on degraded lands required many technical trials in the late 1970s, and foresters looked to improve their expertise with imported species. By 1985, around 500,000 total hectares of new forests had been planted, a rapid expansion over previous decades (Hoàng Hòe 1996).

Important support for expanded reforestation came from international aid, and some of the earliest donor-funded development projects in the post-war era concentrated on forestry. Starting in the late 1970s, the World Food Program (PAM) targeted denuded areas of the Vietnamese countryside, particularly the northern coastal plains and the uplands around Hanoi (MOF 1991b). The PAM money, nearly US$160 million in total, was used to buy seedlings for planting by both households and SFEs, as well as for infrastructure and subsidies. PAM funded nearly 450,000 hectares of forest plantations in the 1980s, channeling money for these efforts through financially troubled SFEs, and helping to keep these increasingly insoluble enterprises open. PAM projects also operated as food-for-work projects for local people during times of severe food shortages, but agricultural insecurity continued to trouble labor availability for afforestation throughout this time (Beresford and Fraser 1992). Households were also primarily involved only as paid labor, without any personal investment in the trees planted, because much of the afforestation was directed at lands remaining in state hands. In a conversation with a foreign forester who had arrived in 1988 to work with PAM funding, he confessed, "In hindsight now, years later, I can see the major problem has been there the whole time, [which is] the land classification and registration problems. That needed to be solved right off the bat."[6]

The first post-*đổi mới* era reforestation project was set up in 1992 and became known as the 327 program, after the decree number given by the Council of Ministers that authorized it.[7] This new policy tied together many of the longstanding goals of environmental rule for upland areas—namely, sedentarization of ethnic minorities, migration of ethnic Vietnamese to less populated highlands, reclassification of protected forests managed by the state, and afforestation—all together. The objectives included:

> (1) To protect remaining forests in areas where ethnic people have either a nomadic or sedentarized life, but still practice shifting cultivation.

> (2) To protect and manage other forests the State uses with other methods such as forest rangers and financial resources.
>
> (3) To reforest barren lands and hills with indigenous tree species and cash crops which have long-term protection effects through agro-forestry approaches.
>
> (4) To link forest protection, reforestation of barren lands and hills with a solution for shifting cultivation as practiced by ethnic people through attracting labor power locally available or from delta provinces to new economic zones which will contribute to the readjustment of population densities among the regions. (Nguyễn Vành et al. 1995)

Program guidelines stated all government levels must mobilize their efforts and financial resources to participate in the goal of reforestation and restoration of five to six million hectares of land, with a minimum 250,000 hectares planted per year (Nguyễn Văn Sản and Gilmour 1999; Sikor 1995).

To implement the program, the central state issued cash payments to provinces for distribution through SFEs and local forest protection departments to encourage the replanting. Much of the funding was spent on buying tree seedlings or paying farmers to protect already forested land once belonging to SFEs.[8] Funds for the program came both from the state and from international donors: the World Food Program contributed up to a quarter of the 327 program budget, with a total of around $200 million spent on the project over five years (World Bank 1998). These donor-supported reforestation projects were a crucial source of support for state budgets in an era of declining Soviet aid; in the early 1990s, 40 percent of the Ministry of Forestry's total budget was designated for the 327 program (Hines 1995).

I visited one SFE in Thanh Hóa province in 1998 in the process of implementing 327 and transitioning to household allocations following the "closing of the door" (*đóng cửa*) to the 8,000 hectares of forest under its control, a result of the Prime Minister's decision to halt the harvesting of timber from certain natural forest areas noted in the previous chapter. Bá Thước SFE was one of hundreds slated for closure to production, and, instead of logging, the nearly one hundred workers and surrounding residents were engaged in the business of replanting the forest and administering funds for the 327 program. The SFE issued one-year contracts, known as green books, for reforestation and protection to households, and families received subsidies of 250,000 VND (US$20) for planting one hectare of forest, plus free seedlings. Families protecting already replanted or regenerating forest received smaller payments of 50,000 VND (US$4) per hectare per year. Despite the focus on households, the SFE managed the entire process in a top-down manner, as they considered themselves the forest "experts," through selecting appropriate species and deciding where and how much to plant, making the 327 program very much a state-managed process of afforestation.[9] Still, the 327 program posted far more successes than had been achieved in previous eras, and all accounts suggest an increase in tree planting did occur; from 1993 to 1998, 1.6 million hectares of forest land were assigned to households for protection, and 1.3 million hectares were planted or rehabilitated (Nguyễn Văn Sản and Gilmour 1999).

SELLING THE FIVE MILLION HECTARE REFORESTATION PROGRAM

Despite serious criticisms of the 327 program's top-down nature, and armed with the overall statistics about a rise in forest cover in hand, in 1998 the government embarked on an even more ambitious nationwide plan, the National Five Million Hectare Reforestation Program (5MHRP, or the 661 program for the decree establishing it), to replant forest so that by 2010 the total forest area of the country would be 14.3 million hectares, equivalent to the forest cover once present in 1943. The idea of reforesting five million hectares had originally been proposed by state officials in the late 1980s, but it took time to translate this vision into a nationally backed program. The closure of natural forests in Prime Minister Kiệt's logging bans, along with the 1991 Forest Law that reclassified forest lands, had added a new urgency to the idea of reforestation, which explained why the 5MHRP emerged when it did (MARD 2011). Unlike the 327 program or PAM projects, the 5MHRP was a national project, not simply one of the Ministry of Forestry, and thus the amounts of funding and support made available was exponentially higher. In the twelve years the 5MHRP ran from 1998–2010, over US$1.5 billion total was spent on the project, coming from the state budget, loans and donor support, and the private sector (MARD 2011; SRV 2011).

The five million hectares goal consisted of interventions in different forest types. Two million hectares were to be forests for protection, while the remaining three million hectares would be for active wood production (Hoàng Hòe 1996). The stated goals of the 5MHRP were three-fold:

> (1) To establish five million ha of new forest in order to increase the forest cover to 43 percent of the land area (thus contributing to environmental

security, reducing natural disasters, increasing water protection, and preserving the source of genes and biological diversity);

(2) To use areas of bare land to create jobs, contributing to the eradication of famine and alleviation of poverty, sedentarization of farming, increasing the income of people in rural parts of the mountainous regions, particularly ethnic people, ensuring political and social stability, national defense and security, especially in the border regions;

(3) To supply wood for industrial purposes, firewood and other forest projects for domestic consumption and the production of export goods, to make forestry an important contributor to socio-economic development in the mountain regions. (Nguyễn Văn Sản and Gilmour 1999, 23)

As the goals of the 5MHRP indicate, multiple objectives long standing in government policy, like the sedentarization of swiddeners and elimination of this practice, were included as policy goals, and the problematization that followed 5MHRP implementation was thus very similar to previous forest approaches. Individual behaviors—like planting agriculture on land the state identified as "forest land," cutting trees the state had claimed, or living in areas classified as watershed protection zones—were the cause of forest degradation, not state forest overuse or misclassification. Other concerns of a social, economic and cultural nature, such as strengthening political stability and national security, were also folded into this so-called "environmental" policy, a reminder that reforestation was less about nature than it was about people.

There were a number of other unstated goals in the 5MHRP as well, at least in the minds of implementing officials. For some, regulation of water for lowland agriculture was a key factor in their support of the project. This was made clear to me in an interview with the head of a local SFE in Hà Tĩnh, who emphasized that the location of forests around Kẻ Gỗ Lake, a large irrigation reservoir, was the biggest consideration when the SFE and local authorities were deciding where to establish new tree plantations and to allocate land under the 5MHRP. This point of view was shared by a Swedish advisor to a forestry project, who had first come to Vietnam in 1983 to assist with the Bãi Bằng paper mill, and who asserted what he thought the real reason behind 5MHRP was: "Water. Water. Water for agriculture. The government is looking ahead thirty to forty years to create a situation where water for lowland agriculture is guaranteed." For other officials, the 5MHRP was a project to help provide political and economic support to state-owned wood industries. The 5MHRP earmarked one million hectares of production forests for supplying paper mills alone, with an additional 500,000 hectares for pressboard, while 80,000 hectares were set aside just for the production of pit props in the Northeast near the important coalmines of Hồng Gai (Hoàng Hòe 1996). All of these were state-controlled industries run by politically powerful people in local provinces. Some officials also focused on the supply of wood to new export industries, as the government had far reaching and ambitious plans to not only be self-sufficient in wood production, to but develop extensive export wood processing industries like garden furniture and plywood, many of which were controlled by party and political patronage machines. The estimated total volume required yearly for these factories was an astounding 11.3 million m^3. After deducting contributions from existing plantations, the net requirement to be supplied by the new plantations was 8.6 million m^3 a year by the year 2010, an enormous amount (MARD 2001a).

Donors engaged with the reforestation program as well, although for different reasons, eventually providing over US$200 million for the project (MARD 2011). State officials deftly managed the process of *interessement*, a key factor in enrolling actors in a network, to bring these donors and NGOs onboard with the reforestation plans. First, state officials often referred to the 5MHRP and other afforestation projects as "regreening" (*phủ xanh*) activities, rather than what they actually were in terms of the largest component of funding, which was "raw materials for pulp mills and chipboard export" actions. In a partnership document between donors and MARD outlining what the 5MHRP would do, environmental considerations were listed first, and a long laundry list of outcomes emphasized:

> The overriding aim of forest establishment under the 5MHRP is to increase forest cover and increase the protection capacity of forests. This capacity serves to protect the environment, by regulating and protecting water sources, preventing surface soil erosion, preventing land slides, reducing floods and drought, protecting key national constructions (e.g. hydro-power stations and irrigation schemes), preventing moving sand, protecting coastlines, and ensuring downstream agricultural production. (MARD 2001a, 7)

In other words, "Forest was presented as the panacea for environmental protection and restoration. In the Vietnamese national imagination, planting trees—regardless of the tree species—has become equivalent to improving soil quality and fertility, increasing water run-off, and enhancing biodiversity" (Clement and Amezaga 2008, 270).[10] Yet in reality, project documents were clear the 5MHRP was not in fact about the environment or biodiversity, at least in terms of where funding would be directed. Five times more of the 5MHRP budget would go to new, fast-growing plantations than would be set aside for the protection of existing special-use forests where natural ecological functions might be conserved. Nonetheless, environmental NGOs, such as the Hanoi-based offices of World Wildlife Fund and Birdlife International, both with biodiversity mandates, were brought on board the 5MHRP through formal partnerships with the state agency carrying out the project, despite the fact their parent organizations had been active campaigners against industrial forestry in other parts of the globe.

When I interviewed foreign aid workers in organizations supporting the 5MHRP about what they thought of the reforestation schemes, I received a variety of answers. Some expressed the general opinion that any tree was a good tree, and the program may have had faults, but overall had a notable goal. Staff stationed in Hanoi also did not have a clear idea of how the planting was happening on the ground once the project began. Notably, donors largely trusted Vietnamese officials' statements about the problem and the solutions; there were no independent early evaluations of the 5MHRP that might have raised red flags about how lands to be reforested were targeted and classified, or how tree species to be planted had been chosen, and donors simply focused their support on the claims for "regreening" the government had produced, believing this would be a positive outcome.[11] Through this strategic discourse, the reforestation projects in Vietnam became a quintessential "boundary object" typical of an actor-network. Alignments around the idea of a regreened, reforested Vietnam brought disparate groups together, none of whom had to individually agree with all aspects of the 5MHRP, but who were engaged in support for different interpretations of the benefits of reforestation, such as rural employment, poverty alleviation, agricultural development, or biodiversity conservation.

VISUALIZATION AND INTERESSEMENT

Visualizations of what the forests of Vietnam once looked like, and what they could be again helped spur the successful *interessement* for tree planting. The government continually promoted the idea that five million hectares was the figure required to bring forest cover back to what it had been in 1943. The obvious questions are: "Why 1943?" and "Where did this understanding of what forest cover was like at that time come from?"

The reforestation figure is loosely based on a 1:2,000,000 scale map showing forest cover in 1943, generated by a French colonial bureaucrat, Paul Maurand, who had long worked at the *Institut des recherches agronomiques et forestières de l'Indochine*, which provided research support to the Forest Service. By the time Maurand published his book *L'Indochine Forestière* (Maurand 1943), the institution had been essentially discontinued under Vichy-Japanese occupation. Maurand's map, published in an appendix to his summary of the state of Indochina's forests, has several interesting aspects (figure 4.02). One, Maurand's map actually showed more than 43 percent forest cover (perhaps as high as over 60 percent in Tonkin, nearly 45 percent in Annam, and only 13 percent in Cochinchina), but most of this land was highly degraded. The map revealed only a few spots of dark green "fairly rich closed canopy" (*forêts denses assez riches*) forests, concentrated in what are now the provinces of Kon Tum, Nghệ An, Thanh Hóa, Hoà Bình, and Bắc Kạn. "Fairly rich open canopy" (*forêts claires assez riches*), "poor closed canopy" (*forêts denses pauvres*) and "poor open canopy-woodlands" (*forêts claires pauvres-savanes boisées*) were the other categories on the Maurand map. However, it is unclear what evidence and data Maurand used to construct these divisions, as no exact definitions were provided.[12] Further, there was a clear political motivation for Maurand's map: he wanted the Forest Service to be reinstituted as an important land manager and regain its place as the prime arbiter of forest knowledge, management, and use in Indochina. (Maurand would continue his advocacy for forestry after the Franco-Vietnam war and into the US-backed South Vietnam regime as well, publishing an optimistic tome in Saigon in 1968 titled *Forest Policy For Consideration After the War in Vietnam.*) One way to achieve his goal of re-bureaucratizing the forest estate would be to visualize the extent of remaining forest, and also its poor condition, which would present an easily grasped argument for the need to retain the French colonial Forest Service.

Maurand's original map, which depicted forest areas by quality, was later replaced with a simplified version in the late 1980s, which, though it had a heading of "forest types," showed Vietnam in 1943 as one vast swath of green, implying a unitary, heavily forested (yet imagined) past that could be recovered (panel one on figure 3.01 previously). Created by the Forest Inventory and Planning Institute in 1985, this visualization was widely replicated, and it served as an important driver in forest plans up to the 5MHRP. The map also served to represent a single national forest estate in Vietnam, showing the unity of the nation after the bitter civil war. The map

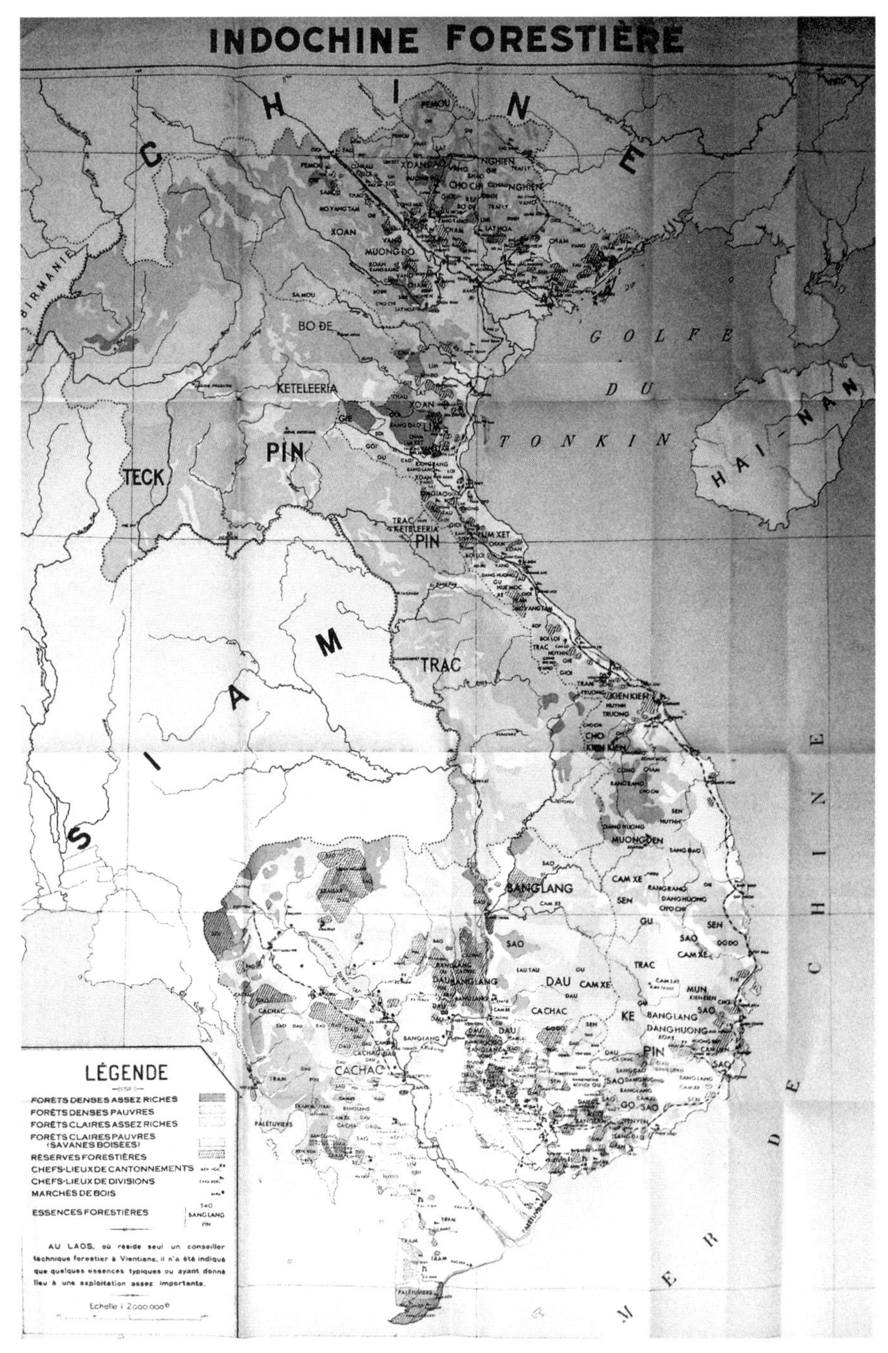

FIGURE 4.02 Paul Maurand's map of forest cover in 1943. Map from *L'Indochine Forestière*, Imprimerie d'Extrême-Orient, Hanoi.

enabled the identification of large areas of land as no longer forests, as these areas had been converted to other uses. They were variously described as "bare hills" and "wastelands," despite that fact that many were in active use. Because they were not the green forest areas the map implied, though, they were considered deficient, and households were mobilized to reclaim and replant them. These new definitions and visioning exercises led directly to policies by the state to expand forest land, and thus in many ways, these maps actually made the object of planning, rather than vice versa.

THE PROBLEMATIZATION OF "BARE HILLS"

As is the case for many projects of environmental rule, the problematization of lands to be afforested preceded detailed knowledge-gathering; rather, maps were made and classifications assigned based on broad and often arbitrary definitions, often centering around the idea of "bare lands" (*đất trống*), "barren hills" (*đồi trọc*), and "wastelands" (*đất hoang*). Bare hills became a political, not an ecological category, and these lands were often treated as valueless, ownerless, abandoned, and in need of rehabilitation.[13] The definition and classification of bare hills in this manner as "places without forest" proposes its own solution—reforest them, whether they have had a previous use or value or not—and demonstrates the ways in which discourse within environmental rule creates its own objects for intervention.

Concerns over bare hills and wastelands can be traced back as far as the pre-colonial era. Many lands falling outside of easily defined properties, such as private rice fields or imperial dominion, were often labeled "wastelands"; these included reclaimed alluvial areas, river banks, pastures, meadows, and the like (Kleinen 2011). Often used by the poorest members of villages and managed as commons, *công điền* (public rice fields) and *công thổ* (other public lands, which could be dry fields, pastures, residences, or forests) occupied significant portions of village territories in the pre-colonial era (Bassford 1987). Once colonial administrators arrived, however, these lands were often the first to be alienated in public auctions or colonial land grabs (Ngô Vĩnh Long 1990). Outside of delta villages, vast stretches of the highlands raised concern as "bare hills" (*mamelons dénudés* or *collines chauves*) (see figure 1.03 previously). For example, geographer Pierre Gourou complained, "All in all the Massif between the Red River and the Mekong is a rugged, poor, and sparsely populated land… Almost the whole of the area is occupied by a difficult relief, too often covered by a worthless savanna; a short distance from Lai Chau [in the northwest] are veritable manless wastes, given over to herds of elephants." He sniffed at these valueless lands, disfavorably comparing them with European landscapes:

> [T]he savannas are almost without economic importance. They provide a very meager pasture, for the *tranh* [Imperata] becomes tough very quickly. There is nothing similar to the Normandy grasslands or the lush Savoy mountain pasturage. . . . The savanna lands are land of low population, often with a density of less than one inhabitant per sq. km. The moors occupy poor lateritic land, which one finds particularly on the edge of the plains of Annam and Tongking. . . . The appearance of these moors recalls somewhat the European heaths . . . but it is only a question of the degradation of the forest by an intensive exploitation; the interruption of this exploitation would probably entail the reappearance of a forest. (Gourou 1940, 490–91)

Bare hills were similarly a target under the early DRV, when the first Five Year Plan from 1960 to 1965 explicitly focused on "the reclamation of wasteland by the people; we must fully utilize the land which still lies waste and fallow, alluviums and new lands formed by the displacement of the foreshore; combine the small-scale clearing of waste land near the villages with that done in the mountain regions by the population coming from the delta" (VWP 1963, 25). For these planners, wasteland development went hand in hand with population redistribution, as it was assumed one reason for the underused bare land in the highlands was the low population densities; newly resettled delta citizens would be the "shock force" to bring these underutilized lands into the remit of the state (VWP 1963, 159).

By the early 1990s, when reforestation projects were in full development, the quantitative classification of "bare hills" took on new urgency. Bare lands were estimated to cover around 40 percent of the country (Ahlback 1995), and, once these areas were semantically defined, they could then be transformed into green replanted forests. But what counted as a "bare hill"? No standard definition was in use and there had been only extremely limited empirical study of these areas (Smith 1993). Attempts to pin down what the term meant were notoriously difficult. Different ministries, departments, and localities used competing definitions; some inventories included the significant portion of the forest estate that was treeless, while other inventories focused on lands outside official forest classifications. The Ministry of Agriculture defined by slope and topography: "Barren land includes (1) steep

TABLE 4.01 Forested lands and bare hills in state-classified forest areas in 1999

Category	Total Area	Actually Forested	"Bare Land"
Special-use forests	2,000,000	930,000	1,070,000
Protection forests	6,000,000	2,800,000	3,200,000
Production forests	12,000,000	5,540,000	6,460,000
Total	20,000,000	9,270,000	10,730,000

Source: Nguyễn Vành et al. 1995.

mountain slopes that have been denuded by human interference, such as shifting cultivation and logging; (2) hilly regions with bush and scrub vegetation, pasture and high grass; and (3) rocky mountains" (Rambo et al. 1996, 21). Still other definitions, particularly those from cadastral departments, focused on the idea of "unused land" (*đất chưa sử dụng*) or "land without an owner" (*vắng chủ*).[14] Such imprecise definitions of barren hills made it difficult to truly assess the extent of these lands, and foreign scientists urged Vietnamese officials to focus more on ideas of loss of biological diversity, productivity, and stability of ecosystem complexes (Lamb 2011); a PAM consultant advised they be considered "unproductive rather than denuded" (Hines 1995, 14). Yet no single definition stuck: "there is essentially neither a clear definition of 'barren hills' nor an ecological understanding of it," reported one study (Nikolic et al. 2008).

Despite these challenges, statistics were generated purporting to show significant areas of the country as bare land (table 4.01), as well as the regional distribution, with the most significant areas located in the Northern mountains and along the Central Coast, often coinciding with areas of excess SFE logging (table 4.02). Even a majority of the more strictly protected special-use forest estate was deemed bare land, rather than biodiverse forest. Overall, of the official "forest lands" identified during the reclassification exercises required by the 1991 Forest Law, the Ministry of Forestry considered half of them degraded and bare.

LOCAL VALUES FOR "BARE HILLS"

The idea that bare hills were essentially valueless was dominant in state discourse over reforestation in Vietnam. Yet the term "bare hills" was deeply misleading, because these lands usually consisted of some sort of ground cover. Numerous local studies in Vietnam have demonstrated many so-

TABLE 4.02 Regional distribution of "bare hills" in 1999

Region	Total Area	Forested Land	"Bare Land"
Northwest	3,065,000	481,000	2,584,000
Northeast	2,302,000	519,000	1,783,000
Red River Delta	318,000	227,000	91,000
North Central Coast	3,166,000	1,427,000	1,739,000
South Central Coast	3,170,000	1,561,000	1,609,000
Central Highlands	4,754,000	3,397,000	1,357,000
Southeast	812,000	457,000	355,000
Mekong Delta	408,000	79,000	329,000

Source: Nguyễn Vành et al. 1995.

called bare hills were actually under different forms of use, such as rain-fed agricultural production; fallowing fields for swidden agriculture; pasturage for grazing buffalo and cattle; and for fuelwood collection (Gayfer and Shanks 1991; Ohlsson et al. 2005). Indeed, as researchers in one study of areas classified as bare hills noted, "None of the interviewed farmers referred to the research sites as 'barren' land, but as 'fields' or, occasionally, 'fallow fields'" (Nikolic et al. 2008, 22).

This was the case around the KGNR as well, where local people exploited "unused" lands and "bare hills" for pasturage and collection of non-timber forest products (NTFPs) (McElwee 2008; 2010). The province classified around 40 percent of the land in the buffer zone of KGNR as bare hills, but in fact the large majority of such lands actually had vegetation, ranging from grass to tall trees, on it, and only 4 percent was considered badly eroded and truly ecologically barren (unpublished district statistics, 2001). Around the KGNR, the vegetated "bare lands" were dominated by waist-high bushes of species found in heavily disturbed forests, including scrubby native wood species and grasses. In 2001, to determine how vegetated and useful these areas actually were, I set up a 50m x 2m transect (0.01 ha) through one "bare hill," which revealed 22 different species and more than 168 individual plants (for details, see McElwee 2009). I used focus groups and interviews with local households to identify which of these plants had local uses, including as fuelwood, medicines, construction materials, or

FIGURE 4.03 Collector of forest products (here, showing a palm leaf used to make conical hats) in "bare hills" around the Kẻ Gỗ Nature Reserve, 2001. Photo by author.

food and fodder, which revealed that *every single species* I had documented on the bare hill plot had a local use, most for subsistence, but some could also be sold for cash.

Women particularly reported higher dependency on bare hills for NTFP collection, as these areas were closer to residences, while men tended to look for products like timber or rattan in deeper forests now enclosed in the KGNR. Women used the bare hills for collection of fuelwood, medicinal plants, and cattle fodder for subsistence, and palm leaves and grasses to sell to make conical hats and brooms; the latter did not fetch much money, but grasses and palms were easy to find, lightweight, and required only a small knife to cut (figure 4.03). These plants could also be found and harvested year-round, which provided an important "bank" of quick cash supply when women had adequate labor to go harvesting. Women reported they used their income from bare hills for household purposes, such as school fees or supplies for children. This was in contrast to the men's money supplied from illegal timber harvesting, which could be raised less often but in larger amounts, and which often went not to the household, but to men's "gambling and drinking funds," as one woman put it.

REGREENING HILLS OR RECLAIMING LAND? ENROLLING HOUSEHOLDS IN TREE-PLANTING

Bare hills became targets of afforestation programs like 327 and the 5MHRP, both of which shared a reliance on the household labor of rural villagers to plant new tree seedlings in designated areas. This set these projects apart from the earlier pre–*đổi mới* projects relying on state workers, such as SFE employees or forest department technical "experts," to do most of the growing and planting, which was inefficient and slow. The newer projects mobilized millions of smallholders to take on the labor and risks of tree planting, in return for presumed rewards, either the financial or subsistence returns from tree products, or, as was the case for the 5MHRP, the potential reward of newly privatized land.

As noted earlier, one challenge for the PAM and 327 programs was that tree planting and land ownership were decoupled; these projects provided either no new lands to households, asking them to plant trees on their existing agricultural lands, or else gave short-term land protection contracts to farmers in return for afforestation labor. It was not until the 5MHRP that revised land laws enabled the transferring of full private land use rights through long-term land tenure certificates (LTCs, or *giấy sở hữu đất*, also known as red books [*sổ đỏ*] after the color of the covers) to participating households in tree planting projects. Both the 5MHRP and previous projects had all supplied the same types of subsidized tree seedlings and provided payments for households to plant the trees. But it was after the 5MHRP included strong land tenure rights that participation by households dramatically surged, suggesting this land component was a crucial driver (Meyfroidt and Lambin 2008b). Yet the emergence of land as an important marker of class and identity, rather than labor as had been the case in the socialist era, set into motion a process of land grabbing and claim jumping that would have profound effects on local social relations.

NATIONAL FOREST LAND ALLOCATION POLICY

Revisions to the national land law beginning in 1988 allowed households to take primary responsibility for agricultural production away from socialist cooperatives, which codified a large-scale process of agricultural decollectivization (Kerkvliet 1995). While officially all land remained in the hands of the state, according to the land law, the state would "entrust" (*giao*) land to households and organizations for long-term use. A 1993 revision to the Land Law also allowed farmers to request and receive land use rights (*quyền*

TABLE 4.03 Rights to natural forest in 1990, for selected provinces

Province	Total Forest (ha)	National SFE Ownership (ha)	Collective or Cooperative Forest (ha)	Household Ownership (ha)	Other state Organizations (Including Local SFEs, Army, etc.)
Lào Cai	154,982	14,470	0	0	140,512
Hà Tĩnh	164,515	138,785	7,389	0	18,341
Đắk Lắk	1,253,032	1,027,139	0	0	225,893

Source: FIPI 1995.

sử dụng đất) for forests in the physical form of LTCs, which were to be issued to households with five fundamental rights—to transfer, exchange, lease, inherit, and use for collateral. People who received a LTC were to manage the land in accordance with the land classification it was under, which for "forest land" usually meant they agreed to replant it with trees or let degraded forest recover. While before 1990, not a single household nationwide had secured land tenure rights to forest lands in the form of an LTC (see Table 4.03), the remaining years of the decade would see a strong expansion of household participation in forestry.

Depending on locality, LTCs were issued for different types of "forest lands," including open access lands, locally managed lands, and those that had previously belonged to the state, such as SFEs dissolving after the logging bans of 1993. What most of these lands had in common, though, was that they were not in fact forested despite being called "forest land." Lands with substantial tree cover often remained in SFE hands or else were transferred to other state entities, like special use and protection forest management boards, which continued to control the largest amount of forest land after the devolution processes (table 4.04). Poorer quality lands were usually the ones made available for allocation, with the hopes that households would take on the burden of reforesting them. According to the closing report on the 5MHRP, 1,037,000 red books for forest land were given out to individual households by the program to 2010, and household participation in tree planting was declared a success (MARD 2011).

LAND ALLOCATION IN THE KGNR

My fieldwork around the KGNR in the early 2000s coincided with 5MHRP money arriving in the district, which enabled me to see the process of forest land allocation up close. Land allocation of agricultural fields during

TABLE 4.04 Forest Land Ownership by Entity, 2010

Organization	Total Forest Area of Ownership (in ha)	Percentage Total Forest Estate
Special use and protection forest management boards	4,318,492	32
Households	3,287,070	25
Commune administration*	2,422,485	18
SFEs and other enterprises for production forests	2,044,252	15
Military	243,689	2
Communities	191,361	1
Other economic organizations (e.g. cooperatives)	91,537	Less than 1
Other groups & organizations	659,935	5

*This is essentially land that is not allocated; it can therefore encompass locally worked out lands, as well as open access lands.
Source: MARD 2011.

decollectivization and the allocation of forestland took two very different paths. For agriculture, by 1990 officials in all villages around the KGNR had started the process to divide productive lands formerly belonging to cooperatives. Depending on the village, land was allocated to households according to a proportionately equal system, but with slightly different formula. In the areas in which I did research, one village used a formula of $100m^2$ of non-irrigated crop fields and $266m^2$ of wet rice fields per person in a household. In another village, the formula was 500 m^2 of wet rice per person and $850m^2$ of garden land, but no non-irrigated land. An explicit goal of the agricultural distribution system was to minimize the number of debates over who got the best spots, and so many fields were parceled out in minute amounts, which explains why most households had on average four to seven individual plots of rice field and two to four plots of non-irrigated dry field. Perhaps as a result of these locally negotiated systems of land allocation, I heard of few conflicts over agricultural land distribution.

But there were important differences in how "forest land" was parceled out, as the egalitarian and locally configured rules guiding agricultural allocation did not apply. First, and perhaps most importantly, forest land was allocated by state employees at the commune and district level, unlike the allocation of agricultural land, in which villages had a say in the process. A team of district officers generally carried out the allocation of forest land, and they based their task upon existing topographical or cadastral maps, which resulted in much confusion. In the case of the allotment of agricultural lands, local officials and farmers had active roles in land assessment and mapping.[15] In addition, forest land was not given to all households, as was the case with agricultural lands. Households wanting to receive forest lands had to be proactive and submit applications to higher authorities, and their applications were evaluated on the "capacity" of the household, usually the labor and capital ability, to reforest the land. Although seedlings were usually given freely by the afforestation projects, it took time, water, and fertilizer to plant them, and labor effort to protect them. Poor families were assumed to be unable to do this, and so wealthier households were able to use tree planting as a way to extend and solidify their rights to lands. This encouraged richer and more well-connected families to dominate the requests for forest allocation, leading to far more unequal landholdings of forest land than had been the case in the allocation of paddy and other agricultural lands.[16] Poorer households were the greatest users of formerly open-access forest and commons lands, and yet they were the least likely to receive LTCs through the 5MHRP. *Not a single household classified as poor* in my survey sample of 104 households had received any allocated forest land from the 5MHRP at the time of my research, while the richest one-third of households had received on average one hectare of forest land. The larger forest plots given to the richer households skewed overall landholdings to a considerable degree; while richer households only had 1.2 times the agricultural landholdings of the poor, they had *21 times* the forest landholdings (McElwee 2009).

The 5MHRP and other reforestation programs also indirectly encouraged people to lay claim to sections of bare hills or other ownerless lands and commons to plant trees and then exclude other users in the hopes a LTC would be forthcoming later after their trees established their claim to land.[17] Over 15 percent of the households I surveyed had staked unofficial land claims to "bare hills" by 2001. Of these households, the richer ones claimed much larger tracts of land, with a mean of 1.6 hectares claimed by the richest one-third of households and 0.35 hectares for medium income households. Only two households classified as poor had claimed lands during my survey, and they had only cleared less than a tenth of a hectare each. Once lands had been claimed by households (even without a formal LTC), the new owners would restrict others from using any forest goods found there; as one neighbor to a land claimant noted in an interview, "three years ago, we used the bare hills quite a lot. Now we don't go at all—because it is being planted with trees from the 327 program and the forest allocation programs, or else it is now being managed as protection forests."

Reports from other areas of Vietnam confirm the land allocation and devolution processes often increased conflicts and exclusionary processes as households competed with former neighbors and kin for new land rights (Sikor and Trần Ngọc Thanh 2007; Trần Ngọc Thanh and Sikor 2006). In interethnic communities, it was often the Kinh households who benefited from land allocation and forest land claim jumping, rather than indigenous ethnic minorities, who were slower to stake out bare hills or other lands for plantations (Castella et al. 2006; Sowerwine 2004; Thulstrup 2014). Conflicts continually erupted over land and grazing rights in many sites of the country post–tree planting, an indication the so-called bare hills and unused lands targeted for reforestation were in fact very much used (Nguyễn Quang Tân 2006). But grazing cattle or collecting grass did not establish firm claims to land: rather, trees did, and so the tree seedlings given out by the 5MHRP and other projects became objects that created new forms of social relations.

But why did villagers willingly alienate village commons or other actively used lands for individual gain? The consistent reinforcement of the household as the new locus of governance in the post–*đổi mới* period created emergent subjectivities and relations to land and trees as belonging to individuals, not collectives. An inevitable outcome of this transfer of authority to individuals has been the exclusion of others.[18] While some localities resisted forest allocation to private individuals for fears of exclusion occurring (Sikor and Đào Minh Trường 2000; Sowerwine 2004), the residents of the KGNR largely saw the process as inevitable. They had witnessed the changes in land tenure brought about by state claims very clearly when the KGNR was established and enforced by state rangers. Thus villagers largely accepted the appropriation of much of the bare hills for forest allocation contracts, fearing they had no recourse otherwise. Village headmen in particular received generous land allocations themselves in all my study villages, due to their access to the political process, and thus they were not likely to protest the enclosure of the commons on their villages' behalf.[19]

Another reason that the exclusion of neighbors from former commonly managed lands was more or less tolerated was the state's emphasis on the environmental benefits of tree planting, which would accrue to the whole nation. Therefore, the tree planting and land appropriation by richer neighbors seemed less selfish than it might have otherwise. Rather, the tree-planting programs allowed households to claim new subjectivities of

FIGURE 4.04 1977 propaganda poster encouraging citizens to "Plant many trees for a beautiful and rich nation." Photo by author.

patriotic citizens who were cooperating with a state vision of a regreened rural sector, rather than what they might have seen as: richer households selfishly marking new stakes to land. The idea of tree planting as moral improvement was widespread, and propaganda and posters encouraged the idea of valiant citizens engaged in forest protection and expansion (see figure 4.04). Newspapers were filled with stories of households who had planted new forests and who were now financially stable, able to send their children to school, and were "progressive" pillars of their communities (Xuân Lạc 2000; Hùng Tráng and Trọng Chàm 2001).

Similarly, those who failed to re-green were pointed to as morally culpable, particularly those citizens who had long been targeted as forest destroyers: ethnic minorities and swidden agriculturalists.[20] On several occasions, I heard district and provincial officials working for reforestation projects admonish ethnic minority households as being unwilling or uneducated enough to know what to do with the subsidized tree seedlings projects were giving out. On one visit to Lào Cai province, the entirely Kinh staff of the district division of forestry had decided many minority households were known to be "lazy" so they avoided giving those households seedlings for reforestation (despite the fact a stated target of the reforestation projects were for poverty alleviation and settlement of minorities). The minority communities who were being discriminated against in Lào Cai had their own views on why afforestation was difficult in their villages. In one Dao minority village I visited in 1999, some families simply said they did not want free seedlings, as they did not have the time or the land to plant them. When they were given seedlings out of the blue anyway, they left them out unplanted, and when the trees died they were yelled at by authorities, who said they would not receive any more in the future. Another man said he wanted to be involved in tree planting, and went to an agroforestry training to learn more, but then got discouraged because the trainer told them they had to fertilize the exotic tree seedlings. "Can you imagine carrying fertilizer up these hills?" he laughed, pointing at his swidden field up a steep slope, and he decided not to participate after all.

OF SUBJECTIVITY AND SPECIES

An additional factor in the new subjectivities and social relations caused by tree planting projects was the tree seedlings themselves. All of the major state afforestation projects relied strongly on a narrow range of species and provenances, especially on eucalyptus and acacia, both Australian in origin (Lê Đình Khả et al. 2003). The focus on these particular species was not only a result of their physical characteristics (namely, being fast growing) but because they were part of a package of technical assistance accompanied by Australian experts. A project on "Seeds of Australian Trees and Domestication of Australian Trees" had provided millions in support for Vietnamese research institutes for seed and field trials as part of a larger $20-million project across Asia (Byrne 2008). Prior to this donor support, improved forest seeds from Italy, France, Holland, India and China and other places had been imported into Vietnam (Trần Xuân Thiệp 1996). But after the Australian project, *Eucalyptus camaldulensis* (*bạch đàn* in Vietnamese), with a provenance from one single site in Petford, Australia, became the dominant tree species used throughout the country.

The Australian project to promote their trees had developed in response to early failures to encourage local households to plant exotic species. In the late 1960s, farmers became alarmed by field trials of eucalyptus in Vĩnh Phúc province, complaining that well water levels near eucalypt plantations had dropped. Further,

> some marshy areas have also become much drier after eucalypt plantations were established; and under plantations there were no grass for cattle grazing. Public opinion against widespread eucalypt plantation was voiced

> and led to disastrous measures. In 1977, the former Bình Trị Thiên Province (now Quảng Bình, Quảng Trị, Thừa Thiên Huế Provinces) acted, and by order of the Provincial authority, over thirty million eucalypt seedlings were killed in nurseries. During 1988–1989 protest was voiced in south Vietnam, under the pretext that eucalypt planting might destroy the environmental equilibrium there. (Trần Xuân Thiệp 1996, 21)

Following this revolt against the exotic species, the Forest Science Institute of Vietnam, a government research center under the Ministry of Forestry, quickly moved to calm fears and establish new trials of other eucalyptus varieties that might be less controversial, and their efforts were funded by the new Australian donor project. This campaign to encourage acceptance of eucalyptus was successful; while in 1980, eucalyptus covered only 5 percent of the area of afforestation, by 1990 eucalyptus was fully 50 percent of all plantations (Fisher and Gordon 2007). In more recent years, a hybrid form of acacia (*Acacia mangium* × *Acacia auriculiformis*, in Vietnamese known as *kèo lái*) has become the second most popular afforestation seedling (Mitlöhner and Sein 2011). Both eucalypts and acacias were planted in monocrop stands in high density, and given this similarity to agricultural cropping, Ministry of Forestry officials even looked into whether vast areas of bare hills could be aerially seeded by dropping tree seeds or saplings in low passes from planes, I was told once. (They could not.)

In fact, these Australian exotics created new forms of conduct: namely, these trees could not grow themselves, nor be aerially seeded, and thus active afforestation was required. In a natural forest, trees can grow spontaneously: a seedling can emerge, given light, water and soil. But in the world of forestry, exotic species require money and labor above natural conditions, and the biological characteristics of the Australian trees were important drivers of afforestation programs. For example, acacia requires dense manual plantings because it is poorly self-pruning (Mitlöhner and Sein 2011). And eucalypts as well as acacias tolerate little competition from surrounding species when they are young, which requires farmers to clear whole stretches of land to plant. Further, both species are sensitive to chemical herbicides, which means hand-weeding is necessary during growth stages, and supplementary fertilizers are often required to help establish the trees in the first one to two years in poor soil. Plantations are then usually clear-cut between six to eight years later. Each of these activities required active, involved, and extensive human labor, which is how households became the prime force in the tree planting campaigns.

Had other species been chosen for reforestation, households would not have been necessary subjects of the campaigns. But regenerating forests naturally with local Vietnamese trees was not considered a viable solution due to the length of time that would be involved, although some foreign consultants had recommended "protection and natural regeneration of some barren lands would be a cost effective method to reestablish forest cover and produce subsistence wood products for local consumption, often with no or minimal opportunity costs" (Hines 1995, 15). Localities received only 1,000,000 VND (US$80) per hectare from the central government for lands under natural regeneration, while those lands to be actively afforested with plantations would receive 2,500,000 VND (US$208) per hectare (MARD, 2001a). Not surprisingly then, most districts chose to convert areas that could have regenerated naturally into plantations. Some observers suggested that focusing on replanting gave job security to Forest Department and SFE staff, who would be needed to grow, sell, and advise households on proper care of seedlings, another reason regeneration was disadvantaged. Others have argued the international pulp and paper industry essentially subsidized a supply of low-cost inputs through donor aid under the 5MHRP, which influenced the choice of species for plantation establishment (Lang 2002); acacias in particular were favored by pulp mills due to higher cellulose content, and acacias ended up being the favored species of the 5MHRP.

Peasant farmers participating in 5MHRP had little choice of which trees to plant, and they accepted the seedlings provided to them from SFEs or extension services, as there were often few private greenhouses in rural areas. Once these seedlings were obtained, any native plants previously found on allocated forest lands were usually completely cleared and uprooted, on advice from extensionists that new seedlings could not tolerate any other competition. On a hot day in May 2001, I watched one family in Village A next to the KGNR prepare their newly allocated land for seedlings of hybrid acacia. The family cut scrubby endemic bushes at the lower stems and prepared the land with hoes and spades, then dug holes carefully a few inches apart to plant the seedlings in pure stands. The family told me their forest plantation was being treated much like their rice paddy field: the land was cleared and cleaned (with burning, if needed, while in other areas, chemical weed killers were used), and seedlings were carefully placed at fixed intervals. The option of letting scrub brush already on the land grow up to become forest again had not been provided as a possibility to this household, they recounted, noting they had been told their lands would be inspected later by 5MHRP officials to ensure the exact number of seedlings they had been given had been planted in the ground.

The questionable environmental benefits of converting natural scrub into plantations were compounded by increased pressure on other forest lands as a result of the 5MHRP. Although the 5MHRP program was meant to re-green the scrubby brush lands surrounding the KGNR, the exclusion of poor households from former bare hills and commons actually ended up pushing people who used to get fuelwood and other forest products further into the KGNR instead, undermining any linking of the reforestation to improved biodiversity for the protected area. Many of the collectors of NTFPs and fuelwood I interviewed said when they were asked to stop collecting in now-privatized bare hill land, they simply went a bit further into the boundaries of the KGNR to continue their activities, particularly to places they knew rangers did not patrol. Additionally, direct replacement of natural forests, not just "bare hills," with plantations has been documented in other provinces of Vietnam, particularly by acacia hybrid plantations in central Vietnam, where a number of pulp and paper mills have a strong presence (Thulstrup et al. 2013).

The lack of protests regarding the use of exotics under 5MHRP was striking, especially considering the earlier dislike of eucalypts in the 1980s. Australian exotics have been strongly critiqued in other countries for being invasive and reducing stream flow near plantations (Kull et al. 2011). Their role in serving as soil anchors and windbreaks have also been reproached; acacias in particular are very vulnerable to blow-downs during tropical storms, and the ways in which they are grown in Vietnam with extensive land clearance for planting has been shown to lead to serious soil erosion (Nambiar et al. 2015). But the lack of protest against these trees in Vietnam was very likely due to the fact they were being actively planted by neighbors, rather than anonymous corporations, as has been the case in other areas of the world (Gerber 2011). In these global conflicts, locals have protested against loss of land to plantations, more than the trees in plantations themselves. In Vietnam, there have been few large corporate plantations initiated and hence some have called this an "absence of land grabs" (Sikor 2012). However, although global capitalist penetration into the plantation sector in Vietnam has been minimal, SFEs continue to control a great deal of land (much of which is now plantation) and have been subject to increasingly loud clashes with local households in recent years where land allocation has been slow (Tô Xuân Phúc et al. 2013). Further, the land claiming that went on in the 1990s and early 2000s by wealthier households can certainly be compared to enclosure movements and land grabs elsewhere, as the outcomes were the same: dispossession of the poor, primarily as a result of classification of their actively used lands as "bare hills."

GENDERED SUBJECTIVITIES

The ways in which the tree-planting projects have also affected gender subjectivities has not been well explored in Vietnam, but much evidence from around the world suggests planted trees are often seen as the domain of men, and actively planting trees may be a way for men to establish control over women's land.[21] Similar patterns can be seen in how Vietnamese women lost de facto rights to bare hills they previously used for NTFP collection, which were then enclosed through the 5MHRP and other projects. When these lands were allocated and trees planted, two interlinked outcomes were visible. One was that the previously used scrub and other plants of the understory, such as broom grasses, were hoed and uprooted for new seedlings. The women thus lost access to valuable naturally occurring plant products, as NTFPs do not generally grow well in the understory of eucalypts or acacias.

Then, when the seedlings were large enough to be sold for timber, the proceeds from the sale of these trees generally reverted to men. I visited the new forest farms of all households in my survey sample that had been allocated land by 5MHRP (thirteen households). In each household I visited, when interviewed about who would control these trees, the answer was: the husband or male in the household. The men decided when to plant, cut, and sell the afforested trees. Women almost never took the initiative to cut down trees, or even branches, from planted forests or gardens, as this was seen as the man's responsibility.

Women's involvement in forestry was often hidden, given these stereotypes and expectations regarding men's roles in afforestation. Women were the main providers of fuelwood for their households, but rarely was fuelwood seen as a "forestry" decision for which females might have been consulted in the implementation of the 5MHRP. Women in focus groups told me some of the worst fuelwood they had ever used came from eucalyptus and acacia trees; both species produced stinky, smoky fires. The favored characteristics that came up most often when discussing fuelwood included wood that cooks well at high temperatures: burns fragrantly; kindles well; has little smoke; leaves charcoal behind after burning; is light to carry; is easy to find; is easy to dry when cut wet; and fetches a high price at the market. In these characteristics, local species ranked highly, while the species introduced by 5MHRP and other projects trailed at the bottom. Yet because fuelwood was seen as a low-importance, "subsistence" activity (despite the fact that nearly every rural area in Vietnam has a viable fuelwood market) and an activity dominated by women, the national tree-planting programs paid no attention to including attention to fuelwood species as part of their aid packages.

REVISITING THE FOREST TRANSITION

By the close of the 5MHRP program in 2010, the government declared it a resounding success, largely on the basis of their own reports of an increase in overall forest cover from 28 percent in 1998 to 39 percent in 2010 (MARD 2011). These reports highlighted which provinces had planted the most trees on the most land (see map 4.01), and end-of-project evaluations trumpeted the targets achieved, in the form of tables on numbers of trees planted, funds expended, and land acreage impacted. While the campaign had fallen short in many areas (particularly in afforestation of production forests), other outcomes were deemed successful, such as an increase in household forest protection contracts above planning targets (see table 4.05). State officials regularly invoked these statistics as an indicator of the positive achievements of their afforestation policies. According to the FAO Forest Resources Assessment reports, issued every five years, Vietnam's forests, which covered 9,363,000 hectares in 1990, expanded to 11,725,000 hectares in 2000, 13,077,000 hectares in 2005, and 13,797,000 in 2010 (FAO 2010), leading many to assert Vietnam has successfully moved from net deforestation to net afforestation during the period of the 5MHRP (Mather 2007; Meyfroidt and Lambin 2010).

But despite the acclaim for the project, there were no on-the-ground independent evaluations of actual results from the 5MHRP. For example, of the 2,416,413 hectares of defensive/protection forest allocated through contracts to households and units for protection, there were no studies available on whether forest protection had actually occurred and what the quality of lands under these contracts were (MARD 2011). Only 2,450,010 hectares of land were claimed to be afforested, short of the 3,000,000-hectare original goal, and the government provided no evidence if these afforested lands still had trees on them at the close of the project. Similarly, regeneration was celebrated as a 28 percent success over targets, but there was no evaluation if the regenerating forests had actually improved physiological structure or function. Not surprisingly, there were no examinations of the social impacts of the project either. While MARD claimed 1,249,602 households

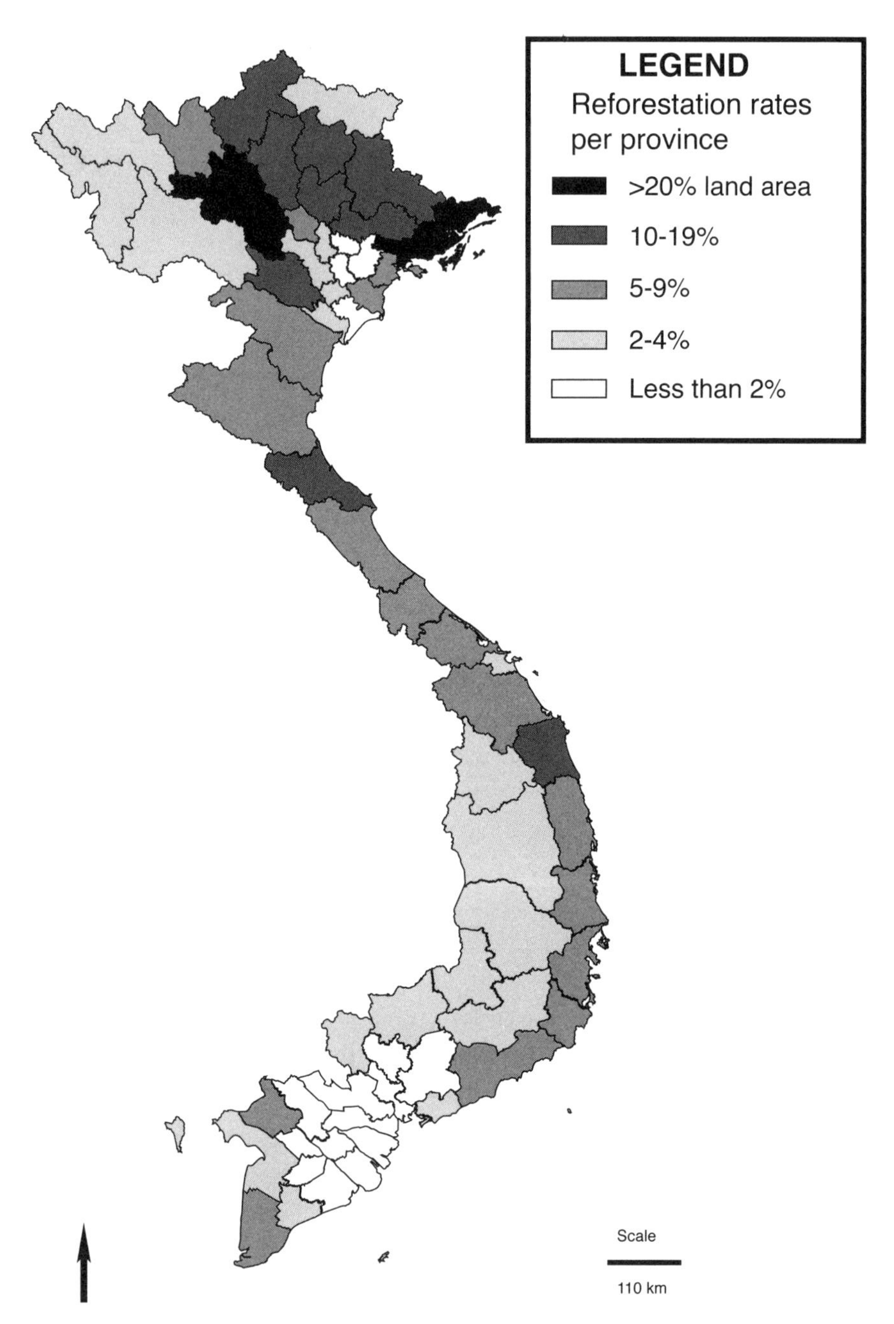

MAP 4.01 Reforestation rates by province under 5 Million Hectare Reforestation Program, 1998–2010. Data from MARD 2011, and base map from Vietnam location map by Uwe Dedering on Wikimedia Commons. Map redrawn by author.

TABLE 4.05 Official results of 5MHRP 1998–2010

Type of Forest Activity	Original Target Objective	Actual Achievement	Percentage of Target Achieved
Protection contracts	2,000,000 ha	2,454,480 ha	120
Natural regeneration	1,000,000 ha	1,283,350 ha	128
Afforestation	3,000,000 ha	2,450,009 ha	84
Of protection and special use forests	1,000,000 ha	898,087 ha	89
Of production forests	2,000,000 ha	1,551,992 ha	76
Of industrial crops	1,000,000 ha	114,767 ha	11

Source: MARD 2011.

had participated in the program in some fashion, of which 484,893 were poor households, no empirical research was conducted to show how these poor households benefited, and how many other poor households might have been harmed from reforestation projects, as was the case in the KGNR.

Given these unclear outcomes and a lack of genuine evaluations of the project, a presentation of results of the 5MHRP by the Ministry of Agriculture to the National Assembly was met with disbelief by several lawmakers, who noted the rosy statistics of successful tree planting targets and raised forest cover did not sync with their local experiences of continued deforestation:

> Ksor Phước, chairman of the National Assembly Ethnic Council, said recently, 'I lived and worked for many years in the Central Highlands where there were many dense forests, but those forests have disappeared. . . . In 1999 there were forests in the area between Đắk Lắk and Đắk Nông Provinces. I can no longer find them, but the report says the forest area has increased. I do not know where else there has been an increase, but in the Central Highlands, a major forest area of the country, I only find forests shrinking. I am doubtful about the reported figure.' (Anon. 2011, 1)

Remotely sensed imagery of afforestation and deforestation throughout the period of the 5MHRP appears to back up these concerns; afforestation was most prominent in Northern and coastal Vietnam, while deforestation remained concentrated in the Central Highlands (Hansen et al. 2013).

Indeed, the most recent satellite data seems to indicate a forest transition may never have happened at all. According to a report in *Science*, from 2000 to 2012, rates of afforestation improved in Vietnam, but there was still a net loss of 664,600 hectares of forest throughout the 5MHRP time period (Hansen et al. 2013). And according to an even more recent satellite analysis, Vietnam's forest cover actually dropped by more than two million hectares from 1990 to 2010, with Landsat data showing Vietnam lost 60,000 hectares a year from 1990 to 2000 and 172,000 a year from 2000 to 2010, the height of the 5MHRP (Kim et al. 2015). (Some of the discrepancies between remotely sensed data sets comes from different definitions of forest; for example, FAO uses 10 percent tree cover as definition of forest, while Kim et al. (2015) defined forest cover "as parcels >1 ha in area and comprising pixels with >30% tree cover.") The government of Vietnam has not clearly specified how it defined "forest" expansion in the 5MHRP evaluations, nor how outcomes were measured (such as by local districts reporting upwards or by satellite and aerial photos), leaving continuing ambiguities behind as to whether or not Vietnam has been afforesting or deforesting, neither or both, in recent years. These ambiguities in definitions are in fact likely deliberate parts of environmental rule, as they allow the networks that brought afforestation into being for multifaceted reasons to also claim multiple outcomes of such projects.

CONCLUSION

Most reports on Vietnam's forest transition are celebratory, and declare the recent tree planting policies a success, based on quantity of trees planted. But forest transition as triumphalism misses much of the process by which these changes happened. While Vietnam has been lauded as a country where afforestation has been driven by smallholders, with an absence of land grabs and no large scale industrial plantations as in other countries (Sikor 2012), the dilemmas raised by afforestation around the KGNR force us to ask: what counts as a land grab? There may be no large multinational corporations kicking people off of their lands for massive pulp plantations, as have been documented for some other countries (Gerber 2011). But the 5MHRP and other tree planting projects embodied the contradictions of environmental rule by putting into place a process by which common lands that used to supply a diversity of forest products have been replanted with exotic trees for timber and pulp production. Poorer households were the greatest users of bare hill land before it was slated for reforestation, yet the least likely to receive this privatized afforestation land, and as such, these local land grabs have operated as a form of enclosure. The outcomes of the reforestation of the bare hills in Vietnam have been similar to enclosure movements elsewhere, where more powerful and wealthier households are able to restrict access to land and economic benefits (Schroeder 1997; Lastarria-Cornhiel 1997; Woodhouse 2003). These outcomes might have been predicted if the tree-planting projects had been recognized early on for what they were: projects of environmental rule that mixed social motivations for policy with environmental justifications for a regreened countryside.

Environmental rule operated quite successfully in the sense that plantations expanded and exotics were introduced with few protests. Households willingly participated in these projects, albeit for their own purposes: not so much to regreen the hills, but to claim land for household use at a time when there was much ambiguity over ownership. The land rush of the 1990s and 2000s is now over, with the frontier largely closed. Those who were able to claim land during this period now have control over it (Mellac 1997; Sandewall et al. 2010); those who lost out will be unable to reclaim possession of new territories barring the dramatic release of SFE and remaining state lands. Had households been asked to sacrifice their agricultural lands for plantations, rather than been encouraged to stake claims on bare hills, it might have been a different story. Horror stories abound about other smallholder plantation projects in mainland Southeast Asia, in which smallholders were encouraged to take out loans for tree plantations on their agricultural fields, which ultimately failed, leaving households in serious debt; these types of experiences were fortunately rare in most areas of Vietnam (Barney 2008).

Environmental rule at this time also differed from the territorialization documented for other tree planting projects around the world, whereby states use afforestation as an "implantation of state power" to acquire control over people and land (Brouwer 1995, 288).[22] In the case of Vietnam, households, not the state, carried out tree planting, although the state was hardly absent from this process given its role in land classification and allocation. Rather than establishing control over the countryside by planting trees itself, the Vietnam state instituted tree planting policies that enrolled individual households in these activities, leading them to new subjectivities as land managers and afforesting armies, roles which were then celebrated with nationalistic rhetoric. Such control was of a different type than physical territorialization, but no less effective.

Processes of land alienation and exclusion necessary to propel the tree-planting campaigns were made possible by the ways in which bare hills were problematized. Classification of lands as "barren" opened up a series of actions—financial, political, and otherwise—that served to transform these areas. Because of an initial lack of information about the use of and dependence on land classified as bare hills, already used lands were misclassified as "unused" and "forest to come," radically restructuring tenurial access to them. While the names implied that these lands were valueless and degraded, in fact, they harbored a number of useful species and were often exploited by people who had no other income alternatives. The lack of information about the vegetation on these lands directly led to priority being placed on reforestation and conversion to plantations, rather than on improving natural regeneration. The visibility of monocrop plantations can be easily contrasted with the invisibility of multiple use lands and NTFPs present in the bare hills all along. This raises the question: what if bare hills had instead been called "regenerating forests"? To what degree did the semantic classification of lands as "bare" contribute to perceptions of valuelessness and to the invisibility of their use? Such questions indicate how important discourse and definitions are to projects of environmental rule.

In ascribing to forests the power to affect multiple types of change (hydrologic, biologic, and otherwise) and stressing the national benefits to nature, project planners enrolled enthusiastic support from multiple audiences. But there is no doubt tree-planting programs operated as a clear form of environmental rule. Across Southeast Asia, similar policies and practices have been enacted, where forest cover expansion has not truly been about "environmental" benefits of a regreened landscape, but other reasons: political patronage; territorialization of lands to keep authority for forest departments (Peluso and Vandergeest 2001; Barr and Sayer 2011); control of rent, subsidies, and other financial incentives (Ascher 1999; Dauvergne 1998); or extension of the state into areas dominated by ethnic minorities and other questionable characters (De Koninck 2000, 2004). The focus of the 5MHRP project on the benefits from "regreening," with the assumption that any forest is a good forest from the point of view of nature, likely kept many from probing more deeply into the possible social impacts of tree planting, such as alienation of lands or the disempowerment of women. These outcomes could have been predicted from the start, had this policy been more clearly identified as a "social project" and subjected to social impact analysis and involvement of specialists in socio-economic issues before implementation.

One conclusion to draw from this chapter is that while Vietnam certainly has lessons to impart to other nations about forest management, these are not about how to make a forest transition (which may never have even occurred), but about how states have been able to successfully enroll both donors and households to fund and provide labor for tree planting efforts, despite the mixed environmental successes that have resulted. Large areas of land now covered with monocropped eucalyptus have not contributed to biodiversity conservation, despite the fact that this goal was a primary reason that several conservation donors supported tree-planting programs. Indeed, the species diversity that long characterized Vietnam's forests, and which was a continual impediment to forest management in the French and DRV eras, is now greatly diminished, replaced with single species plantations now composing the near majority of the country's so-called "forests." To even refer to these new plantations of eucalyptus and acacia as "forest" has required an ontological shift, when they might just as easily be called tree farms.

Understanding the dynamics by which environmental rule has operated in the guise of tree planting is particularly important given new global initiatives to increase tree planting and forest cover for carbon sequestration, which I explore in the following final chapter. Unless the social and environmental consequences of smallholder reforestation programs are better understood, more areas of Vietnam may be under pressure to convert natural vegetation or so-called "bare hills" into plantations of exotic fast growing trees for financial benefits from carbon or other ecosystem services in the future. The consequences for existing land users, and particularly poor households and women, will likely be negative should this plantation expansion continue to ignore their needs. These outcomes would be all the more ironic considering Vietnam's privatization and tree planting programs have been promoted to donors as "regreening" of degraded lands and the better protection of forests for local use and management, while many of the outcomes thus far seem to have had the opposite effect.

5 Calculating Carbon and Ecosystem Services

New Regimes of Environmental Rule for Forests

LIKE MOST DEVELOPING COUNTRIES, VIETNAM HAS RECENTLY become involved in new approaches to environmental policy that purport to rely less on the state and more on private actors to value nature and provide the capital needed for preservation. These policies range widely in focus and scope, and have been called "neoliberal," "decentralized," "hybrid," or "market-based" approaches (Agrawal and Lemos 2007). What they share in common is the use of economic incentives to value "environmental services" like carbon sequestration or water flow in the hopes that the market will provide a more efficient, less expensive way to arrest degrading activities than traditional state command and control policies. Payments for environmental services (PES), which transfers funding from users of ecosystem processes to those who provide soil, water, and forest conservation, is one of the most well known of these new approaches. Another emergent policy would provide funding from international carbon buyers to forest-conserving communities, known as "Reduced Emissions from Deforestation and Degradation" (REDD+).[1] While these schemes are relatively new and untested, particularly in tropical developing countries, in a rapid period of time Vietnam has become an enthusiastic supporter of, and is sometimes pointed to as a global leader on, both PES and REDD+.

Yet, for much of Vietnam's history as a socialist nation guided by a Communist Party, free markets were anathema. For example, in 1958 North Vietnam formed a "Committee for the Reform of Private Commerce and Industry" that reported to the Politburo, and President Ho Chi Minh consistently warned citizens about the errors of "individualism" (*chủ nghĩa cá nhân*) and capitalism (*chủ nghĩa tư bản*) (Yvon 2008). Even after the strict anti-market stance of the socialist era gave way to open economic policy under *đổi mới*, the Center for International Forestry Research (CIFOR) declared in 2005 that PES would be a "non-starter" in Vietnam due to its long history of top-down, state-led environmental management and young and untested system of private property rights (Wunder et al. 2005). Nonetheless, by 2010 Vietnam was being praised as the first Southeast Asian country to have a formal nationwide policy supporting PES.

This transition to newer market-based policy approaches will not spell the end of the projects of environmental rule outlined in previous chapters. In fact, in many ways, environmental rule will and can continue to thrive under so-called decentralized or economic-based approaches to environmental interventions as well. Vietnam provides the ideal case study of this: the country has claimed the mantel of market-based policy for environmental services, without making many fundamental changes in the ways forest management is actually practiced. Indeed, officials tasked with administering PES spoke to me of their activities as a way to recapture new forms of capital, particularly from industrial actors and international donors, in an era of shrinking budgets, but which will be used for long-standing state goals, such as continuing to focus on stopping swidden agriculture, or providing supplies of water to economically important downstream users. Once again, despite new wrapping, an ostensible focus on ecology and environmental services is being used to further social and economic policy goals.

PES has expanded rapidly in the last five years, and this chapter traces the development and implementation of this policy, drawing attention to the many similarities to previous forms of environmental rule. While PES is purportedly in place to assist in valuing ecosystem services, among participating localities there is little understanding of what these services are, and almost no measuring or monitoring of them. Rather, government officials continue to pay the most attention to a metric they have long focused on: acreage of forest. And just as the idea of "environmental services" has been used to value the same aspects of nature that have long been of interest to the state, the idea of "payments" has similarly involved reversion to long-standing policy. Neither PES nor REDD+ operate as markets in any shape or fashion; payment levels to households for forest protection are set by the state, similar to the previous payment-for-protection plans first pioneered under reforestation projects twenty years ago. The state continues to be intimately involved in all aspects of PES and REDD+, from problematization to intervention.

One could even argue that the state's hand in forest management is being strengthened in Vietnam via so-called market-based conservation through the continued salience of governmental forest "expertise" and direction, which has been given new value by PES and REDD+. Like previous approaches to environmental rule, these more diffuse forms have also required problems to be identified (such as what environmental services were and where they could be quantified in value for payments) and knowledge to be compiled through novel techniques of calculation (such as how much carbon was in which trees). These activities have led to new visualizations and mapping practices: namely, a focus on watersheds and carbon, which have served to reprioritize valuations of forests as those that serve the production of hydropower or international carbon markets. Much of the knowledge work that has been necessary to build PES and REDD+ has been devoted to "making things the same" (MacKenzie 2009): that is, making complex ecological processes amenable to a numerical valuation, such as payment per hectare, that can be used across the country. These valuations also serve ontological work: to reconceptualize forests not as timber or trees on land (as previous eras have defined them) but as stocks of carbon or flows of water, and to reconceptualize households not as forest labor or land managers, but as new subjects—namely "providers" of environmental services.

The new focus on environmental services also shows clearly how donors and international institutions become entangled in environmental rule, rather than simply the state, as previous chapters have focused on. Networks of international donors, NGOs, local officials, economists, policymakers, and foresters have served to push PES and REDD+ forward, all for different reasons, and these networks have circulated the use of market-based approaches with multiple justifications. Organizations with wildly different mandates, from timber supply companies to international biodiversity NGOs, have joined together in tenuous networks to support the idea that environmental services like water or carbon should be paid for, although these actors have differed on how this should occur. Other networks have emerged as well, such as a consortium of powerful local actors led by state-owned rubber companies, who have taken advantage of tree and land reclassifications required by REDD+ to push to have rubber considered a carbon-rich "tree" species, rather than an agricultural cash crop, as it had long been classified.

By looking at PES and REDD+ as continuing forms of environmental rule, we can see how new social practices and forms of agency have been created. New subjectivities have begun to be formed surrounding who was worthy of ecosystem protection payments and what should be done with them. Unexpectedly, households who have participated in PES have engaged in their own forms of resistance, insisting that payments that come to households for forest protection need to be equal in size. This idea that neighbors should not get different amounts of cash based on different environmental services or quality of forest protected violates most principles of market efficiency, the entire reason for PES in the first place, but for locals, market efficiency violated norms of social equity. Thus, these new market and valuation schemes have served to energize local communities to protest against remuneration that was considered unequal, taking advantage of the new transactional relationship between households and the state, and creating new forms of environmental subjectivities once more.

MAKING "MARKET" ENVIRONMENTAL POLICY IN VIETNAM

The idea of using capital raised from market forces for the protection of biodiversity was first broached as early as 1996, when a conference was held in Hanoi to explore the topic "Creating revenues from biodiversity in order to conserve it" (MoSTE 1996). However, at that time the focus was on tangible products that could be derived from plants and animals in protected areas, such as medicinals from bioprospecting, and the idea of creating "virtual" commodities like carbon or other environmental services was not raised. Ecosystem services approaches were briefly debated in the late 1980s when the massive Hòa Bình hydropower plant began to operate northwest of Hanoi (Nguyễn Quang Tân 2011). At that time, the state electricity company and officials considered compensating local people for protecting slopes upland from the reservoir, but the eventual resolution was to use reforestation projects, like the 5MHRP, to pay for afforestation. The idea that water and energy users, rather than the government, ought to foot the bill for reforestation and protection was not yet in the realm of possibility, given the strong role of the central government in forestry in the 1980s.

It was only later in the 2000s that officials and researchers first began to hear of the concept of payments for environmental services (*chi trả dịch vụ môi trường*). PES were originally proposed as a voluntary economic transaction between a buyer and a seller of a environmental service in which some sort of provisioning (forest protection leading to water conservation, for example) was offered in exchange for some type of conditional payment (Wunder 2005). Costa Rica's 1996 Forest Law is widely recognized as the first such attempt to use PES at a national level in a developing country (Chomitz et al. 1999). Donor projects and conferences introduced PES to Vietnam, and the first formal attempt to define ecosystem services and to pay households for their upkeep began in 2002; this was funded by the International Fund for Agricultural Development, Swedish International Development Agency, and World Agroforestry Center (ICRAF). This project, titled "Rewarding Upland

Poor for Environmental Services" (RUPES), was designed to introduce policy for PES in several countries in Asia through pilot schemes connecting service providers in upland forested areas with downstream water buyers (Leimona et al. 2008). Although RUPES primarily operated in Indonesia, Nepal, and the Philippines, a small office in Hanoi brought the findings of these pilots to Vietnam, shared through reports and workshops, and also funded small research projects to see if PES could be viable in Vietnam. These efforts were aided by the high-level connections of one of the ICRAF staff, whose husband would later be the Vietnam ambassador to the United States, and RUPES was able to circulate the new PES ideas to policymakers who mattered. In response to the RUPES work and global attention to PES, MARD began conducting research on the economic valuation of environmental services starting in 2004, indicating a clear pathway from donors to government advocacy for PES.

RUPES's focus on the uplands was a direct consequence of the long-standing assumptions about the connections between downstream water flow and upstream forest cover, particularly in the northern mountains of Vietnam. Yet in many ways the RUPES project did not so much introduce new forms of knowledge about the commodification of ecosystem services as it relied on previously produced tropes of upland degradation. Indeed, documents from the RUPES project mirrored government assertions over Malthusian degradation scenarios in the uplands, such as extensive deforestation, soil erosion, poverty, and the need for outside interventions to slow these processes. As an example, the dangers of swidden agriculture that had been the dominant narrative about the uplands for decades were repeated in an early RUPES scoping report, which stated as fact the idea that "In Vietnam, most ethnic minority people live in uplands, practicing slash and burn cultivation. Slash and burn cultivation has been one of the main causes of forest loss and land degradation. Furthermore, it also makes people management by the government difficult" (Bùi Dũng Thể et al. 2004, 20). This conflation of control of unruly ethnic peoples by the state with the need to manage highland environments comes straight from the long histories of environmental rule in the uplands from the past one hundred years, although the promotion of a market mechanism as an intervention for this problem was new.

While RUPES and other donor projects sparked intellectual discussion about PES, it was not until a later project, the Asia Regional Biodiversity Conservation Program (ARBCP), funded by the US Agency for International Development (USAID) and implemented by Winrock International, that households were actually paid for ecosystem protection. The ARBCP began in 2007 in Lâm Đồng province, a forested region of the Central Highlands, and included several traditional "conservation and development" approaches, including providing non-timber tree seedlings to poor households to increase their income. But it is the PES components developed to link forest conservation in upland watersheds to fees paid by downstream users that the ARBCP became most known for. Project officials had ambitious goals for their PES component, stating that by harnessing markets effectively, Vietnam would be able to

> help stimulate local economic growth, public–private partnerships for biodiversity friendly economic activities, and increase financial support for environmental protection. . . . Such a policy could reduce the costs of water and power production for urban areas, provide additional income for thousands of poor families living in forest areas, and provide funds for meeting Vietnam's National Forest Management and Biodiversity Conservation Action policies. (USAID 2009, 1)

Like many other projects of environmental rule that had come before, the ARBCP problematized the situation before actually producing knowledge about forest types and users in Lâm Đồng. The key assumption in inception documents was that people cut down trees because they were poor, thus ethnic minority communities living near forested lands were particularly targeted by the project. Like RUPES, documents from the ARBCP project referred to the destructiveness of swidden agriculture, implying that many upland residents were ignorant of the true value of forests (ARBCP 2009). Only after the project had been fully funded and cooperation agreements signed with local provincial authorities did ARBCP set about producing new knowledge about the value of ecosystem services in Lâm Đồng. However, rather than working to assess the priorities and values of the ethnic minority communities that were to be the targets of intervention, the project produced knowledge about and visualizations for the *users* of ecosystem services. These users were located downstream, and technical consultants were hired to conduct economic studies of the value of hydrological services provided by Lâm Đồng's major watersheds. These experts drew up maps of the province, with administrative boundaries effaced, replaced with watershed boundaries instead (see map 5.01). Areas that fell into a watershed would be able to participate in PES; those that did not would be unable to benefit. Significantly, the major watersheds in Lâm Đồng mapped in this visualization exercise were already the sites of interventions by the state and private investors in

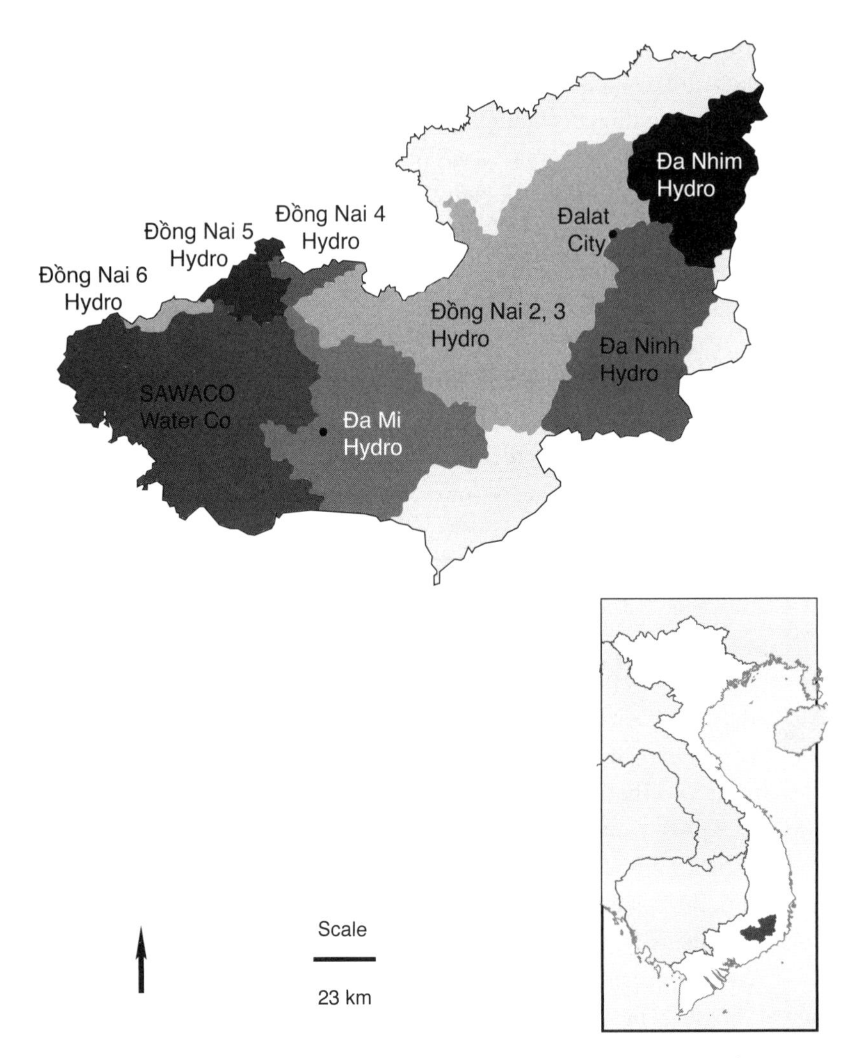

MAP 5.01 Map of Lâm Đồng Province divided by watersheds for distribution of payments for environmental services. Base map from Department of Agriculture and Rural Development, Lâm Đồng Province, 2009. Redrawn by author.

the form of hydroelectric plants, and each major watershed was linked to a downstream generation station in the map.

From these results that watersheds were providing significant benefits to hydroelectric companies and to water supply companies, the ARBCP project proposed a fixed user fee for water and energy companies and its staff worked with Lâm Đồng provincial officials to lay out a legal framework for assessing these fees to organizations outside the province's boundaries, namely water users downstream in Vietnam's richest urban area, Ho Chi Minh City, and hydroelectric generating companies elsewhere in southern Vietnam. However, these efforts could not be implemented without national policy assistance as they involved entities outside Lâm Đồng province, so the ARBCP turned its attention to Hanoi.

ARBCP staff strongly promoted the idea of PES to national-level officials in MARD, MONRE, and the prime minister's office, engaging them in the *interessement* that characterizes network formation. At approximately the same time as the ARBCP project began, MARD was leading a process to design and formulate an official PES policy for Vietnam, including a detailed review of international PES experiences, and ARBCP stepped in to help, such as by sponsoring workshops to promote experiences of PES elsewhere in Asia. More significantly, the project funded an overseas fieldtrip for key Vietnamese officials to PES sites in the US; this study tour included stops looking at ecotourism fees in Hawaii, salmon protection fees in Oregon, and water use in New York City, where they visited the PES scheme linking city water users to watershed protection in the Hudson River Valley and the Catskills. In an interview, a high-ranking MoNRE official who went on the trip noted "This study tour was considered a very useful exercise for Vietnam as those who participated were able to write a report about what they had seen and then organize a workshop to share information from the tour. Information about the tour was later submitted to the Prime Minister, who then approved the legal documents for PES to be implemented elsewhere."[2] Other international experiences on PES were sought out; for example, the United Nations provided technical assistance for a consultant from the Pubic Utilities Company of Heredia, Costa Rica, to visit Vietnam and discuss his country's experience with PES for watershed services.

These enrollment efforts by ARBCP and other donors paid off. In 2008, the prime minister approved a PES policy framework, titled "On The Pilot Policy On Payments for Forest Environment Services," which MARD had developed with the help of ARBCP.[3] This decision set up two PES pilot projects for a two-year basis: one in Lâm Đồng to be assisted by ARBCP, and one in Sơn La province to be assisted by a German aid agency, to be replicated elsewhere in the future if effective. After two years of local payments, the pilots were considered to be successes, and, in late 2010, a new nation-wide policy was passed: Decree 99 NĐ-CP, "On the Policy for Payment for Forest Environmental Services," which states that "Organizations and individuals benefiting from forest environmental services must pay for forest environmental services to forest owners of forest that creates the supplied services."

FIGURE 5.01 Sign in Lâm Đồng Province promoting the benefits of ecosystem services, 2011. Photo by author.

Decree 99 indicated that five types of PES payments in Vietnam would be legal:

> 1) payments for land protection, such as soil erosion; 2) payments for watershed protection and water regulation; 3) carbon sequestration payments; 4) landscape and biodiversity protection payments for tourism purposes; and 5) payments to protect the spawning grounds and source of seed for aquaculture.[4] (MARD 2010a)

At the time this decree passed, only seven other developing countries (namely Mexico, Costa Rica, Ecuador, El Salvador, Brazil, South Africa, and China) had national PES legislation, making Vietnam the first country in Southeast Asia to implement this policy (McElwee et al. 2014). The decree was widely represented as a win-win situation, whereby both economic and environmental goals could be achieved (see figure 5.01).

Vietnam's statute states that many PES fees will be mandatory and that buyers will include hydropower companies, domestic water supply companies, industrial facilities that use water, tourist companies, and others to be determined.[5] Rather than use a market to set PES payment levels, a fixed user fee was instituted nationwide: 20 VND per kilowatt hour (kwh) (US$0.0013/kwh) on energy generated from hydroelectric plants and 40VND per m^3 (US$0.0025/m^3) domestic water consumed in participating urban areas.

Tourism companies depending on some sort of environmental service would also be assessed 1 to 2 percent of their total related revenues. From the beginning of the policy to the second quarter of 2014, more than $157 million USD had been raised by PES. I was told by the head of the agency in charge of PES fees that this was the first policy in Vietnam to raise so much money in such a short period of time, and currently PES is providing about 90 percent of the funding for forest protection work in Vietnam.[6] Significantly, much of this money for forest protection was essentially coming from clients of another bureaucratic ministry, the Ministry of Industry and Trade (MOIT), which oversees hydropower and energy production, and which had long been a more powerful and better funded office than MARD. To a large degree, officials I spoke with in MARD thought of Decree 99 as a "win" over MOIT, and a chance to build a stronger base of power within the government for MARD, hereby demonstrating the unspoken goals that often undergird environmental rule.

NETWORKED INTERESTS, DISPARATE GOALS

As government PES activities ramped up, so too have donor-funded PES or PES-like projects expanded. As of 2012, at least nineteen different donor projects that have labeled themselves PES or market-based were operating or in planning stages (Phạm Thị Thu Thủy et al. 2013; Tô Xuân Phúc et al. 2012). Sponsors have included large donors like the World Bank and United Nations Development Program (UNDP) who are working on forest carbon; conservation organizations like Birdlife International and the World Wildlife Fund, who are funding biodiversity valuation and marine and mangrove protection through user fees and sustainable certification programs; and poverty NGOs like Care International and the Dutch organization SNV who are supporting payments for watershed protection. PES has thus provided an opportunity for a number of actors to advocate for funding of their particular concerns, whether they are wildlife preservation, poverty reduction, or water conservation, under the broad rubric of "environmental services," and, as such, PES serves as a clear exemplar of a boundary object (Barnaud and Antona 2014). For donors involved in these projects, the concept of "environmental services" can encompass multiple representations of and objects in nature (animals, or water, or trees), as can the idea of "payments" (as cash, contracts, markets, or subsidies). PES policies thus take on multiple meanings and enroll other actors through the porousness of the concept; not everyone has to agree completely on what PES actually is.

The relationships that have sustained PES as a boundary object can be explained by understanding the interconnected networks that brought PES to Vietnam. In this way, we can see the arrival of PES not as some sort of outside "neoliberal" imposition of an economic valuation approach, as some have argued (MacDonald and Corson 2012), but as a relational idea that has become incorporated into multiple agendas and thus internalized or indigenized in the course of translation. Several international NGOs and donors have asserted that while they brought ideas and resources to state actors in the discussions of PES, they did not have direct influence on the development of Decree 99 itself (Phạm Thị Thu Thủy et al. 2014). But these donors did have influence on the knowledge networks that solidified interest in the new concepts of "ecosystem services" and "valuations." Just as the network of organizations overlapped and intertwined with each other, a few key people worked for both donor and government projects on early PES approaches, like academic economists Đặng Thanh Hà and Bùi Dũng Thể (working for ARBCP and RUPES, respectively); forester Vũ Tấn Phương of the Research Center for Forest Ecology and Environment, who has done carbon assessments for many donor projects; and government advisors Phạm Xuân Phương, former vice head of the legislative department of MARD, Nguyễn Chí Thanh, former head of the Forest Inventory and Planning Institute in Saigon, and Nguyễn Tuấn Phú, former head of the economics department for the Government Office (*Văn phòng Chính phủ*) of the Prime Minister. By moving between donor PES projects and pilots and state agencies looking into the development of a national PES policy, these individuals served as important nodes linking authority and knowledge to action and intervention. The intersections of the donor network, with their multiple goals for PES, and state officials who quickly praised the PES approach, did however raise disparate interests. While donors stressed the involvement of new actors, like NGOs, businesses, and poor households as "service providers," government actors have often had very different perspectives and have emphasized the continuing importance of state agencies, expertise, and management.

"SOCIALIZING" STATE CONTROL OF RESOURCES

In interviews, government officials expressed the most support for PES with reference to the money that can be raised for state goals of forest conservation, not the decentralization aspects or need for enrollment of new actors in forest management. For government agencies, PES promised a new way to fund their ministries' core missions: retaining and expanding forest cover and increasing supplies of forest income and goods. This emphasis is best captured in the text of Decision 380 initiating the two pilot projects, which articulates the explicit goal "to socialize the forestry sector, gradually establishing sustainable economic basis for protecting the environment and ecosystems, improving quality of service provision, especially ensuring water supply for electricity production, for clean water production, and ecotourism business activities." The neoliberal practice of moving from state-provided public services to the private sector is called "socialization" (*xã hội hóa*) in Vietnam, rather than privatization; it is intended to mean that "society" must bear greater burdens than the state in service provision. What this means in reality is that in the new market-oriented era, many former state services are now funded through additional out-of-pocket fees and contributions paid by individual citizens, and cost cutting measures have been taken in many sectors, such as increasing class sizes in schools.[7]

Socialization of environmental protection was similarly imagined as a way to fund core activities through user fees, rather than central state budgets. The expressed hope for Decree 99 is that it will enable the funding of forest conservation activities without the need for top-down government transfers; the days of large centralized funding projects like the 5MHRP appear to be over. The vice-director of the General Directorate of Forestry (also known as VNFOREST, located under MARD) noted in a meeting in late 2011 that his office hopes to only supply around 25 percent of the budget for forest management to lower-level state entities (such as National Parks, Forest Protection Management Boards for watershed/protection forest, or SFEs) in the future, and the remaining 75 percent of budgets will have to be raised through creative means such as PES, entrance fees, or other approaches.[8] Yet this decentralization of fund-raising is not meant to diminish the role of the state in directing forest policy; indeed, the former deputy prime minister (and now head of the National Assembly) has stated clearly "The policy of PES will create the conditions for the state to manage forest resources and forest land more tightly, and at the same time allow people to participate in the protection of forest more effectively" (Nguyễn Sinh Hùng 2010). Others echoed the idea that PES is good for state control; a local official in Hoà Bình province told me that despite a long dislike of markets in Vietnam's political history, PES is actually an approach that is "fitting with the Party" (*gần Đảng*) in the sense that the Party's roots have always been in the rural peasantry, and PES, to this official, was a way of getting back to prioritizing this sector.[9]

Tellingly, I have never heard an official in Vietnam refer to PES as "market-based" in more than fifty interviews on the subject and countless meetings over the past five years; the idea of a market (*thị trường*) is completely submerged in Vietnam's adoption of the practice of PES. State officials readily acknowledge that pricing of PES, which is set at a fixed standard in the national law, is hardly a market practice. In fact, "it's like the subsidy era," as one participant at a PES meeting in 2014 described it, by which he meant the high point of socialism when the state was involved in nearly all aspects of life. Some NGO projects that have tried to develop more market-like interventions (for example, certification of sustainable timber or seafood, or direct payments between a user of environmental services like a brewery and upstream communities) are in fact not considered PES by government officials, as I found out when I tried to write a report for VNFOREST arguing that these NGO projects were in fact good examples of PES. I was told they were not; PES in Vietnam is only officially recognized when it involves regulated entities paying into state-managed funds the fixed environmental service fee outlined in Decree 99. Nothing else will be counted as PES, at least for the time being: PES is what the state says it is, and that means it is not a market.

PROBLEMATIZING AND DEFINING ENVIRONMENTAL SERVICES

Unlike interventions in previous eras of environmental rule, PES as an intervention came first, before problematization or knowledge construction. PES was introduced by donors and embraced by state officials as a tool for raising funds from new sources: the idea of payments was clear. But what about environmental services? What was the problem PES was supposed to fix? In other settings, the primary reason to use PES is to value an undervalued asset, such as forests that are threatened by land conversion, whereby a PES payment could increase the value of retaining the forested lands. But in the case of Vietnam, problematization has been incomplete: there has been insufficient attention to what an environmental service actually is, and how these services have been threatened by degradation relating to a lack of valuation has almost never been discussed by officials.

Globally, attention to ecosystem services expanded rapidly after the publication of the *Millennium Ecosystem Assessment*, which identified a number of services that were undervalued in national accounts, including services for supporting ecosystems, provisioning food and other goods, regulating climate and carbon cycles, and cultural services (MEA 2005). However, the commodification of environmental services in order to turn them into easily valued goods for which payments can be measured has required complex interactions between scientists, bureaucrats, and the financial sector (Bumpus 2011; Lansing 2011). Much of the complexity relates to different definitions and concepts that underlay ideas about "ecosystems" and "services," as these terms move from their origins in "scientific" disciplines to applications in the real world that require attention to social and cultural differences and power (Forsyth 2005; Kull et al. 2015). Many critical scholars have focused on the problem of PES creating "fictitious commodities," as Polanyi (1944) first termed the creation of goods for a market, arguing that these represent a nefarious attempt to capitalize previously public or free goods from nature (Brockington 2011; Kosoy and Corbera 2010). However, knowledge production about ecosystem services has far outpaced the production of valuations and payments and markets. Indeed, there has been a rapid acceptance of the idea that ecosystem services are real and measurable as a standard "fact," rather than a complicated and contested field of knowledge (Barnaud and Antona 2014; Kull et al. 2015). In this way, the concept of "ecosystem services," and particularly ideas that upland forest conservation provides downstream water "services" of value, has become a type of new "environmental orthodoxy" (Forsyth 2002).

The difficulties of simply measuring many ecosystem services, let alone defining them satisfactorily to multiple audiences, has led policymakers who have tried to implement PES globally to focus on a narrow range of services, primarily water provision (Stanton et al. 2010). Indeed, there is an element of predetermination here: certain ecosystem services lend themselves to centralized planning and explanations of environmental cause-and-effect, namely forests and water supplies, given long-standing theories about forests as sponges. Moreover, large-scale water supplies ideally should be provided over extensive land areas in an entire watershed, thereby strengthening justifications for state-led management of these services, even in the name of market policy. Therefore concepts such as "watershed" have become a simplified unit of scale that fits ideas of ecosystem services, even though such definitions are not universal and can be highly contested (Cohen 2012).

Similar challenges have befallen the implementation of PES in Vietnam. Intervention (payments) came first: only secondarily did problematization and knowledge gathering about environmental services happen. In the two provinces that first tested PES from 2008 to 2010, the pilot sites did not begin their policy implementation by assessing where deforestation and land cover change was happening within the province, nor did they define the drivers of degradation of ecosystem provisioning in any systematic way. Rather, problematization was driven by the already-determined goal of collecting payments from some sort of ecosystem service user; thus, the policy identified the most easily assessable service that could be paid for by a user, not by where ecosystem services were most threatened. In Lâm Đồng province, although the PES project did not make initial assessments of deforestation rates or drivers, my own interviews with provincial officials indicated that deforestation rates over the past five years have been about 5 percent per year from the conversion of forests to coffee, from illegal logging within state forests by timber gangs (often working on contracts and run from outside the province), and from development projects like hydropower and golf courses. The province is also Vietnam's top exporter of tea, vegetables, and flowers, the latter two mostly grown in greenhouses by corporations and wealthy individuals, who buy or lease forest land, often through shell companies. Yet these "users" of ecosystem services were either powerful politically or else illegally using resources, and their use of ecosystems was largely destructive, which did not lend itself to asking for user payments.

Instead, rather than working backwards from a threatened ecosystem, the policy focused on an easily defined, non-destructive "user": a downstream consumer of water or energy. This predetermined "problem" led provincial officials in both pilot sites to focus on hydrological services and soil erosion prevention for hydroelectric reservoirs. The ARBCP project paid for external expert evaluations of water ecosystem services, and one study used the Soil and Water Assessment Tool (SWAT) model to simulate the watersheds in the project area and the amount of water flow that might be expected downstream into hydropower reservoirs and barrages given different land use scenarios.[10] The SWAT analysis gave a range of estimates of stream flow downriver from different land uses, which economists then used to estimate the monetary value of regular water flow and reduced soil loss (as sediments that flow into hydropower reservoirs shorten their lifespan) to the 160 MW Đa Nhim hydroelectric plant, built by South Vietnam in the 1960s.[11] The suggested values were that each hectare of forest provided US$14.64 in water regulation and US$54.43 in reduction of sediment per year. This valuation was complemented by another economist's study of willingness to pay among water users in urban areas of Ho Chi Minh City, as at least one partial watershed for the Đồng Nai river which supplied the city with domestic drinking water fell in Lâm Đồng province (ARBCP 2009). Other environmental services, like biodiversity, were judged to be too difficult to assess or compensate directly, so no valuations were attempted. Carbon was an additional environmental service indicated in Decree 99, but the general consensus among policy makers I spoke with was that carbon will only be an environmental service that international users will pay for, so valuations have not been attempted; instead, emergent REDD+ policies driven by international donors and users will fund these services (discussed further below).

From these two reports, the ARBCP recommended that economic valuation of the local environmental services be worth at least 20 VND/kwh generated from hydroelectric plants and 40VND/m^3 assessed from water consumed in participating urban areas. These rates were approved by MARD, and not only did MARD allow them to be assessed for the two initial pilot sites, these figures were later adopted for the entire country in Decree 99, despite the fact that they had been suggested based on very geographically limited data specific to one part of one province.[12] In sum, the entire PES environmental and economic structure, rather than being set by a market, was determined primarily by limited assessments of ability of users to finance payments, not on the opportunity costs of service providers (that is, whether or not a payment for forest protection was more lucrative than conversion of forests to coffee or tea plantations or some other land use). This problematization that focused on users rather than providers proved a later sticking point when local communities became involved in payments, as discussed below.

When Decree 99 was adopted for all of Vietnam, technically all provinces should have been required to implement the policy: environmental services, defined broadly, can be found nearly everywhere, even in urban areas (such as trees that provide shade and climate regulation). But only around half of the provinces have moved to set up PES payments, and this has not been based on those provinces that have forest cover or other "natural" areas, but on provinces that have hydroelectric generating plants located within or near their borders. With this smaller subset of participating provinces in mind, they set about assessing ecosystem services by focusing on forest cover as an easy visualization of presumed services.[13] Expertise was deemed essential to these exercises: private consultancy firms (many with ties to MARD) were hired to use GIS and remote sensing to make new maps of watershed basins—primarily focused on elevation and slope—to assess how many hectares of forest in a given watershed could be claimed by each province in order to receive PES funds (Phạm Văn Duẩn and Phùng Văn Khoa 2013). However, this focus has been contested somewhat by hydroelectric companies themselves, who have pointed out that while technically an entire watershed may provide some regulation of water flow, it is generally the forests that are nearest to a reservoir that provide the most important water and soil regulating properties. But the national data sets produced after Decree 99 were not made on the basis of actual water flow into reservoirs, and instead focused on general topography and forest cover alone (MARD 2010b).

The end result of this mapping exercise was that 4.1 million hectares of forest land in watersheds were designated as lands for which PES could be paid as of 2014, accounting for around a quarter of the total forest land area in the country (map 5.02). Forests not in watersheds—which includes mangrove forests, coastal wetlands, and Melaleuca forests in the low-lying Mekong Delta—have not been valued for PES, though they undoubtedly provide ecosystem services as well. Given that the mapping of forest lands in upland watersheds has been mostly about slope and topography rather than forest quality, one end result is that PES fees are being collected for non-forested "forest" lands in watersheds (including many so-called bare hills), about which hydropower companies have expressed reservations. Plantation forests as well as natural forests in these watershed maps also receive PES money. While officials in Thừa Thiên Huế province noted in an interview with me that the large majority of their forest lands are no longer natural, thanks to plantation establishment under 5MHRP, and given rotations as short as three years, they were unsure if these trees would count as providing an "environmental service" to receive payments or not, nationally MARD has accepted that they are.[14] To date, several provinces have received PES money for exotic plantations; this has included payments to rubber companies in Bình Phước province and acacia plantations in Đồng Nai province, although it is not clear what services these areas are providing, as hydrological regulation from monocrop planted forests is significantly lower than from natural forest areas (Fox et al. 2012).

If identification of environmental services has been predetermined by the interventions that targeted specific users and narrowly defined components of nature, resulting in inconsistent application of the law, then monitoring of services has been even more uneven. In general, provinces have evaluated the environmental effectiveness of PES policy through assessments of change in forest cover and the number of forest law violations, which are gathered by annual reports of the Forest Protection Department as required by law anyway (but without specific data on PES implementation areas per se). The ease of visualization of forest cover loss was one major reason why it was usually taken as a proxy indicator of "ecosystem services" in general. For example, in Cát Tiền National Park, one of the largest and best known parks in Vietnam which is receiving PES money, this financing has not been used to monitor biodiversity levels, because in the words of the vice-director of the park, "It

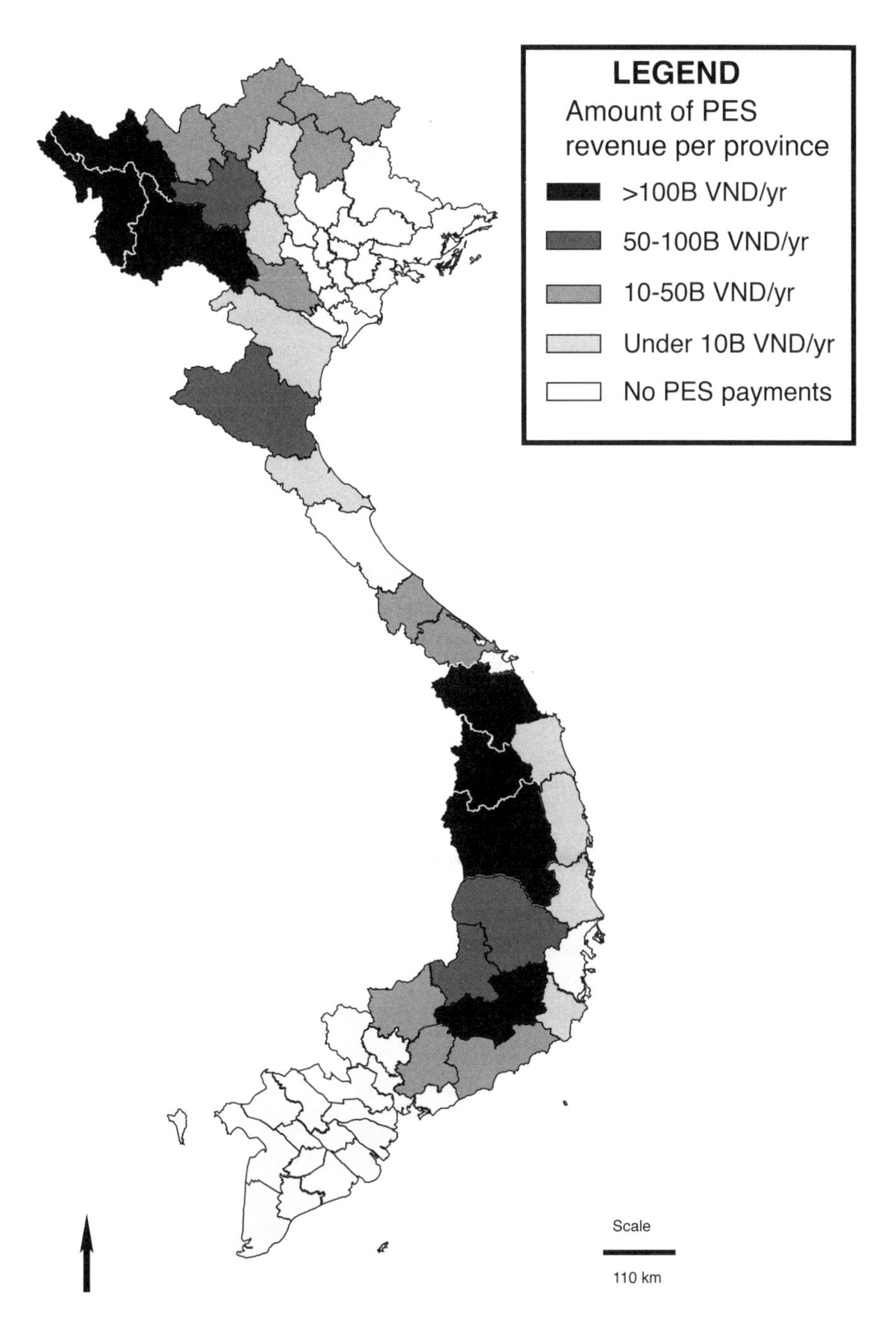

MAP 5.02 Revenue raised from payments for environmental services by province in 2013. Data from MARD 2014, and base map from Vietnam location map by Uwe Dedering on Wikimedia Commons. Map redrawn by author.

is too difficult to set up monitoring of animal numbers; but changes in trees and land are easy to see."[15] MARD has not required recipients of PES money to put in place specific monitoring of hydrological flow, the level of soil erosion, or the change of water level and water quality in reservoirs in relation to the area and quality of forests where PES is implemented, primarily because few provinces have the technical ability to do so. Only one province has actually attempted to measure and monitor new environmental services: ARBCP funded Lâm Đồng to procure monitoring equipment to estimate water runoff and soil erosion in the watershed of Đa Nhim reservoir. But after only two years of operation, the equipment stopped working, and was then disassembled and put in storage.

INTERVENTIONS: COLLECTING PAYMENTS AND SUPERVISING PROVIDERS

PES payments have been assessed for several years now in many provinces, and for over five years in the two original pilot provinces of Sơn La and Lâm Đồng. How these payments are then transferred to so-called service providers gives another opportunity to see environmental rule in action. Just as the intervention came first, before ecosystem services had been defined, so too were recipients of PES money assessed only after payments had been collected. Indeed, in many provinces, PES funds are sitting in a state budget account and have not been passed to any service providers, because they have not yet been defined and identified.

State and donor-led knowledge production has been crucial to defining certain types of forest users as worthy of attention, while others are not. In the two original pilot sites, because they did not identify key drivers of deforestation initially, the ARBCP suggested targeting of payments from PES fees based mostly on subjective criteria of poorer, ethnic minority households. However, in my own conversations with forestry officials in Lâm Đồng province in 2009, when they were first working out the PES policy, they were of the mind that state organizations could be paid out of the fund potentially, not just individual households. Most of the forested land in the province fell under government management, such as in national parks and nature reserves, forest watershed protection boards, and SFEs, and officials argued that these organizations, rather than individual households, had a stronger claim to PES funding. Other provinces made similar arguments in their PES implementation. Officials at Bù Gia Mập National Park in Bình Phước province suggested that they should get to keep PES money to build new guard stations.[16] In a meeting with Lào Cai provincial officials in 2014, the head of the Forest Protection and Development Fund (FPDF) stated:

> The average household receipt from PES here is only 200,000 VND (US$10)—so what impact is that going to have on livelihoods? If we at the FPDF were able to keep all 20 billion of PES moneys collected, we could have a lot of equipment, staff, and other needs met. But when we have to distribute it to households it's just too little to make a difference. Households can go out and work in the wage labor market for 100,000VND a day! So what is 20,000VND once a month going to mean to them?[17]

In discussions with the original authors of Decree 99, they emphasized that PES monies were not supposed to substitute for traditional funds received for forest management, and that every effort should be taken to ensure PES funds went to households rather than state entities, although enthusiasm for this has varied by province. MARD has tended to side with these arguments, rather than letting SFEs and other state management boards have complete control of PES money. The compromise that appears to be holding for now is that the central and the provincial Forest Protection and Development Funds can keep 10 percent of collected PES fees for management purposes, as can state forest owners. (This means in some provinces, only 80 percent of total PES fees collected go to households because the fees go through two layers of state land-owner bureaucracy first.)

The focus on households as the main recipients of PES money has required the identification of which ones might be paid for forest protection, and which ones have less secure claims. Many provinces have chosen to use existing land tenure certificates as claims to PES funding; that is, if households have a red book for forest land that falls in a watershed receiving PES money, they should have a claim. As noted in the previous chapter, however, many poorer households never received red books for forest land during land allocation projects of the 1990s, particularly ethnic minorities with swidden lands or households using common bare hills, and land tenure holdings were often skewed toward wealthier households. Given this, PES funding runs the risk of being directed to households that are already receiving benefits from possessing larger areas of landholdings.[18]

To identify these households, individual provinces have had to come up with new types of visualizations: not only where watershed basins were, but who owned the land within these basins. For example, Lào Cai province hired a consultant group to use SPOT5 aerial photos to determine watershed areas, and then devised a map in ArcView at 1:10,000 scale, which local districts and communes then used to classify individual land owners against land tenure certificates on file with their offices—a time consuming and cumbersome process, given that each individual owner had on average less than 0.2 hectares. Sơn La province ran into similar difficulties, in that 52,000 individual households who owned red books had to be identified and mapped by watershed. In Thừa Thiên Huế province, the slowness of mapping land claims and inaccuracies between official registries and what people have claimed on the ground has meant that no owners of land have been identified as PES recipients yet, and the 12 billion VND (US$550,000) in funds the province has raised in the past three years of PES has yet to pay a single household.

In other areas with low rates of household ownership of forest lands, authorities have left it to state forest owners to make arrangements with local households for forest protection activities, rather than relying on red books as proof of land tenure rights. This is the approach that was taken in Lâm Đồng province, where the bulk of PES money is transferred to individual households living near state-owned forest lands on yearly contracts. Households who agree to participate in PES sign legal statements that they will not engage in deforestation, including occupying forest land or cutting trees for swiddens, hunting, exploiting NTFPs, debarking or ringing trees to kill them, or starting forest fires. (These contracts are almost exactly similar to the solidarity pledges asked of citizens around protected areas in the 1990s, as noted in chapter 3.) PES recipients also must state that they will regularly patrol and protect the forest lands under their supervision, and many have to file patrol schedules with the state forest owners they contract with as proof. Depending on the type of state forest owners and the amounts of land these owners control, protection contracts have ranged from ten to thirty hectares per household. Some villages have chosen to receive PES money collectively, and patrols are done in groups and PES money divided up in equal shares. A total of around 8,000 households have participated in these contracts since the beginning of PES in Lâm Đồng.

As of early 2014, the total number of households involved in receiving PES money nationwide is around 350,000. However, each participating household receives widely variable amounts of money, despite the fact that uniform payment levels have been set for user fees nationwide. This discrepancy in payment rates comes from the fact that households have different amounts of forest land to protect (ranging from less than a hectare in many areas to thirty or more hectares in some protection contracts) and

TABLE 5.01 Average PES payment rates for one-hectare of forest in participating provinces

Province	Total PES Forest Area (hectares)	Number of Hydropower Plants Paying PES to the Province	Payment Rate Per Hectare Per Year (in VND)		
			Highest	Lowest	Average
Lai Châu	434,404	5	382,630	302,837	342,734
Hoà Bình	74,013	5	N/A	N/A	134,234
Lâm Đồng	321,718	14	360,000	325,000	342,500
Gia Lai	493,579	33	265,000	200,000	232,500
Kon Tum	251,346	11	411,000	88,000	249,500
Bình Định	206,609	0	604,690	1,828	303,259
Lào Cai	116,903	23	307,000	38,000	172,500
Quảng Nam	184,568	18	353,000	60,000	206,500
Nghệ An	61,683	8	345,000	236,000	290,500
Bình Thuận	52,084	4	120,000	50,000	85,000
Thanh Hóa	43,015	1	70,000	68,000	69,000
Bình Phước	43,380	2	N/A	N/A	180,000
Điện Biên	242,279	7	256,000	826	128,413
Đắc Nông	214,876	9	223,794	121,139	172,466
Cao Bằng	79,451	12	N/A	N/A	38,000
Hà Tĩnh	25,624	3	463,000	17,000	240,000
Đồng Nai	153,555	0	57,000	8,000	32,500

Source: McElwee and Nguyễn Chí Thanh 2015, based on data from VNFF.

that payments rates per hectare vary by province and even by watershed. Generally provinces take all the PES money they have collected from user fees for the year, and then divide this amount by the number of recipients in the province or subwatershed. More densely populated provinces therefore have smaller amounts of payments, and provinces with only one or two hydropower plants receive less money than those with ten or more. Further, if energy or water use fluctuates from year to year, so too do received PES funds, and therefore payments vary as well. (The range of payments in different provinces can be seen in table 5.01). This unevenness has resulted in some households that I have interviewed having received less than US$10 a year from PES, while others have received more than US$500 a year. The largest payments have been in areas where households are on protection contracts to patrol large areas of state forest land, while households whose claims to PES money rely on red book ownership have in general much smaller areas to protect.

Another factor influencing funding amounts have been attempts by provinces to set different levels of payments for different types of forests, using a "coefficient K" to determine financing schedules. The idea of K-coefficients was included in Decree 99 based on the recommendation of economic experts who noted that there should be some attempt at introducing conditionality and efficiency to PES payments. That is, better protected and densely stocked forests should receive more money than poorly protected and poorer quality forest: such are the tenets of economic valuation that PES originated from. To implement this conditionality, MARD recommended a complicated formula that coefficient K should have 4 sub-coefficients, including:

> K1, based on forest status and quality (K=1 for rich forest down to K=0.9 for poor and regenerating forest);
>
> K2, based on the official classification of the forest (K=1 for special-use and protection forests, and K=0.9 for production forests);
>
> K3, based on the origin of forests (K = 1 natural forests down to plantations at K = 0.8);
>
> K4, based on the level of difficulty in forest protection (K=1 for very difficult, K=0.9 for less difficult).[19]

Provinces found these calculations challenging for many reasons, including the poor maps of forest cover and land ownership that did not allow for detailed assessments, as well as the fact that tens of thousands of small forest owners would need to be individually assessed for their payment rates using these coefficients. In Sơn La province, officials tried in the first year of PES policy to implement their own interpretation of these coefficients, deciding that lands classified as "natural" protection forest would be assigned a coefficient (K=1), which meant that around US$7 per hectare per year would be paid, while plantation forest under protection would be (K=0.9), equivalent to US$6.4 per hectare per year, natural production forest would be (K=0.6) with payment around US$4.2 per hectare per year, and plantation forest under active production would be assigned (K=0.5), or US$3.5 per hectare per year. But once forest officials tried to calculate individual payment rates for the nearly 52,000 red book holders in their province, many of which had more than one plot of land in different classification categories, they found themselves overwhelmed with work, and decided to abandon the idea and simply use one payment rate for every hectare across the board.

In Lâm Đồng, similar coefficients were used to reward higher amounts based on quality of forest cover and degree of management, but protests from households ended up changing the payments. In the first years of the pilot, households did not understand why neighbors were being paid different rates for having invested the same amount of time and labor in protection; some households fell in a watershed with higher payment rates, while other neighbors might have contracts for less land. The unequal payments resulted in resentment and even vandalism toward those who were benefiting the most. In the second year, these complaints and protests eventually resulted in officials calculating payment rates in a simple fashion, based on dividing the total amount of the provincial PES fund each year by the total number of participating households. In 2010, this amounted to around 280,000 VND (US$13) per hectare in payment, and in 2011 it was 400,000 VND (US$19). Protests and requests to use K=1 for all households due to concerns about equity have also been made in Quảng Nam and Nghệ An provinces (McElwee and Nguyen Chi Thanh 2015).

In interviews to assess why the protests happened, households noted that they preferred to see more even rates of payments between forest areas, as they did not understand why they might have gotten less money than a neighbor for having invested the same amount of time and labor in protection. Households emphasized that their work in PES was actually based on *labor*, rather than the state's idea of land area and quality or type of environmental service. For households, they primarily went out to their contracted forest areas on set time schedules, walking around forest edges to inspect it, and they often described their actions not in terms of "protection" but in terms of "effort." Just as jobs of similar "effort" in the old SFE and cooperative era were paid similar wages, households in PES projects wanted equal payments for similar labor. For these households, the value of forest was not measured in terms of its abstract qualities, such as water flow rates, carbon content, or boundaries of watersheds, but in the very tangible and calculable time spent patrolling it.

While PES was supposed to instill new support for forest protection in participating households, who now received cash representing new values for these forests, for many households participating in PES, the project was actually quite similar to previous forest protection and reforestation payments they might have received under the 5MHRP program. In a survey of 150 households in PES participating areas of Sơn La and Lâm Đồng provinces that I conducted in 2011, only 34 percent of survey respondents reported that they had heard the term PES, while the rest had not or did not know. Of those who had heard of PES, half thought it was a government program for forest protection. Only a handful of respondents knew that it was a program that primarily received payments from environmental service users. Most households thought that the PES payments that they had been receiving were direct government payments, and they were not aware that payments had been made by hydropower or water companies to the provincial government, which then disbursed these monies. One household I interviewed that was receiving money to help patrol Cát Tiền National Park told me that PES was "money from the state (*nhà nước*) given to the community to help us ethnic minority compatriots." PES payments were often seen as yet another of state charity-type programs, which made it difficult to connect the idea of conditionality to the payments.[20]

Not surprisingly, given this disconnect to conditionality, fully 25 percent of households in our survey reported having done nothing differently in terms of land use after receiving PES funds, indicating that many households had not changed either their activities nor subjectivities as a result of PES. The difference in our sample between those who had protection contracts as the basis for PES money, and those who had red books, was interesting. Those who had red books were more likely to have stopped using their forests for fuelwood and to have made more effort to regenerate these areas, while those on protection contracts focused more on organizing patrols and stopping outsiders from entering into forest areas, and reporting forest fires to authorities more actively (see McElwee et al. 2014) (figure 5.02). But for the quarter of households that had received PES money yet done nothing, it seemed the new subjectivities that financial payments were supposed to engender had not yet filtered down to them.

Yet even though many PES participants were rather ambivalent about what they were supposed to do in terms of protection activities, some had engaged in active attempts to rethink the terms of payments, at least in the sites that had protested K-coefficients. In those areas, the transactional nature of the payments appears to have emboldened participants to request changes, and to see themselves as partners of the state, rather than simply

FIGURE 5.02 A local forest patrol around Bi Đúp Núi Bà National Park, funded by payments for environmental services funds, 2011. Photo by author.

recipients of largess. Previously, as noted in chapter 3, the relationship between state forest owners and nearby households was usually a one-way obligation: households were to follow regulations, not burn forests, or encroach on them, or do anything deemed illegal. What PES contracts have meant is that state forest owners now have obligations toward households as well: to bring PES payments on time, to listen to their complaints, and discuss forest work with them. Particularly in Lâm Đồng, households and state forest officials who were interviewed all noted that PES had offered opportunities at the very least for the two sides to meet more often, resulting in better, less conflictual relationships. Officials at Bi Dúp Núi Bà National Park believed that there had been far fewer cases of illegal logging and arson after PES implementation, which they attributed to "better feelings" between the two sides.[21]

It remains to be seen if payments will encourage households to assert other claims and rights as part of PES participation, such as claims to more secure individual land tenure in areas where yearly PES contracts, rather than red books, are used. It also is unclear how PES will recalibrate relations between households and the state in terms of who should bear the most burden for forest protection, particularly in cases where deforestation was not caused by the service providers themselves, but by outsiders, ranging from households who sold or had lands taken from them for development, forests areas that were encroached upon by agricultural or other enterprises, and areas that are under threat by illegal loggers. In each case, the legal system is not yet clear who has authority to report and request alleviation of these conflicts: the households receiving PES, or the state forest ranger department. As noted in chapter 3, illegal loggers encompass a range of actors, from locals cutting a few trees now and then to extra-local characters with ties to national crime syndicates, and PES incentives are likely not going to be helpful in encouraging households to confront these outside gangs, given the violence by which they often operate, particularly since PES-funded patrols are provided with neither weapons nor specific training in how to confront illegal loggers.

REDD+ AS EMERGENT FOREST POLICY

The rise of PES as a policy mechanism to protect forests has in more recent years dovetailed with concerns about slowing or halting land use generated carbon emissions as part of climate mitigation efforts. In 2007, signatories to the UN Framework Convention on Climate Change agreed that paying countries for so-called "avoided deforestation" would be a significant impetus to slow global forest loss, and negotiations have been underway since then on policies for REDD+. The fundamental premise of REDD+ is that if households and governments are given payments that equal or exceed what could be raised by cutting down forests, then forests will be better protected and carbon conserved (Corbera et al. 2010). While debate still continues at the global level as to the exact mechanisms by which REDD+ will operate, a number of regional and national projects, primarily implemented by bilateral development donors and NGOs, have begun pilot projects to prepare countries for REDD+ implementation in the future. These "REDD+ readiness" programs include the World Bank Forest Carbon Partnership Facility (FCPF) and the United Nations' REDD Readiness project (UN-REDD), both of which have begun operations in Vietnam.

While REDD+ projects are in their very early stages, and are far less well developed than PES, they share in common an interest in using financial incentives to protect forest ecosystems. At the same time, the emerging regulatory apparatuses that are associated with REDD+ also look a lot like familiar forms of environmental rule, as new problems must be defined (such as identifying deforestation that can be halted through economic incentives) and new forms of knowledge must be produced (such as where forests with high carbon are and what peoples can be paid to protect them). Both of these processes have been strongly influenced by actor-networks of donors, scientists, and government officials engaged in discussions of where, for whom, and how REDD+ financing will be directed. With new problems and new knowledge generated, subsequently new forms of conduct and subjectivity are supposed to follow (such as the transfer of money or other incentives and presumably the replacement of deforesting practices with conserving ones among recipients). Yet unlike PES, REDD+ has been unable to follow through to create interventions in forest communities, and one reason for this is the lack of clarity over what kind of knowledge is needed to turn forests into carbon commodities, and households into participating carbon providers. Indeed, there is some evidence that the ontological uncertainties about what constitutes a forest or a tree that have arisen in the context of discussions about REDD+ may be contributing to forest conversion and deforestation, the exact processes REDD+ is supposed to halt.

PROBLEMATIZING DEFORESTATION DRIVERS, IDENTIFYING FOREST SUBJECTS

Vietnam is estimated to have in the range of 700–1,600 million tons of carbon (Mt C) stored in existing forests and soils, and an average of 10–15 Mt C equivalent emitted yearly from land use, making it a low-to-moderate-level country in terms of global impacts (Hoàng Minh Hà et al. 2010). Since soon after the 2007 UN meeting in which REDD+ was introduced, donors and state ministries have been in discussion about how Vietnam might participate (in Vietnamese, REDD+ is called *Chương trình giảm phát thải khí nhà kính từ mất rừng và suy thoái rừng*). A new national REDD+ steering committee was established by the government in early 2011, coordinated by MARD, and a special Vietnam REDD+ office was established within VNFOREST. In the summer of 2012, the Government of Vietnam approved a National REDD+ Action Plan for 2011–20, which encourages the development of provincial action plans, pilot projects, and legal frameworks.[22] As of 2014, seventeen different pilots were underway in different provinces to publicize REDD+, do carbon baseline measurements, and other activities (Phạm Thị Thu Thủy et al. 2012). Many of the forest sites in Vietnam that were previously prioritized for biodiversity conservation under new forest classifications in the 1990s are now being prioritized for these carbon emission reductions.[23]

Like PES schemes before them, REDD+ interventions bear little to no resemblance to a market. While eventually there may be hope that any confirmed emissions reductions achieved through activities on the ground could be sold in an international carbon market, right now, the REDD+ pilots operate as donor development projects on short-term intervals and with external funding routed through Vietnamese government partners. Further, the interventions proposed by REDD+ projects look a lot like previous Vietnamese state policies towards forests. For example, a "green growth" emissions reduction project aimed at the north central region of the country will be funded with $37 million in financing from the World Bank Forest Carbon fund (with the hopes that nearly twice this amount can be recouped in future years by selling emissions reductions of ten or more Mt C). Six main interventions have been proposed by the project for the next five years: developing provincial REDD+ action plans; increasing forest allocation and planting of trees on "bare hills"; sustainable forest management in SFEs; conversion of "unused" land to industrial crops; payments for environmental services; and forest law enforcement and governance (MARD 2014). In other words, despite the attention to REDD+ as a new tool of market efficiency and incentives, many of the actual on-the-ground activities continue a long line of traditional policies for forest management. Indeed, the above list of interventions reads like a summary of the policies that this book has analyzed as tools of environmental rule: reforestation of bare hills, increased use of forest rangers, and attempts to manage SFEs for sustainable lumber production chief among them.

ONTOLOGICAL DILEMMAS: MAKING CARBON REAL

One type of knowledge production that has been necessary for REDD+ has involved making forest carbon in diverse areas comparable to one another through the identification of tree types that store carbon and the development of a formula to determine payment rates for protection of carbon-rich forests. Such practices are occurring elsewhere as many diverse actors come together to form carbon markets, which require the turning of immaterial ideas into material objects (such as emissions reductions certificates or carbon offsets) (Bumpus 2011). In Vietnam, state foresters, many of whom spent the 1990s worried about being put out of jobs by the decentralization of forest management that was taking place at the time, now have new cachet and demands for their expertise in the development of these statistical formula to calculate forest quality and carbon content. Consultants have been hired to create lists of tree allometric equations, which are an expression of the amount of carbon likely to be found in particular species, based on average stand density, wood volume, wood density, bark-to-wood ratio, and other factors (UN-REDD Vietnam 2013). Such equations fit the definition proposed by Michel Callon of "*dispositifs de calcul*," or calculative mechanisms, that are often obscure and opaque, but which are required to turn an intangible idea into actual material good (Callon 2009; Callon & Muniesa 2005).[24] The replacement of ideas of actual physical forests with representation of forests in the form of allometric equations and payment types exemplifies these mechanisms, and the role they play in enabling new regimes of practices under environmental rule. In this imagining, the calculative mechanisms authorize the transformation of forests from ecological groupings of trees into stocks of carbon. Such calculations can become real circulating objects, with practical material consequences. If certain trees are represented as more valued due to higher carbon content, those trees may be favored for management and replantation above others.

These calculative mechanisms also raised ontological questions: what is a tree? If a tree is simply a mass of woody material made of carbon, then many species that had long been classified as agricultural crops in Vietnam, such as coffee, cashew or rubber, potentially become new actors in REDD+ plans. Indeed, households interviewed in our local sites raised these questions themselves: once told of what carbon was, several households suggested that they might be eligible to receive carbon payments for coffee trees, the very planting of which had driven figures on deforestation that made Vietnam a target for international REDD+ projects.

Evidence suggests that the introduction of REDD+ may have produced new ideas about carbon materiality that have led to new forms of "trees" being officially recognized. Several local provinces in areas where rubber grows well have been pushing to have natural forest areas, particularly watershed forests, reclassified so that they can be converted to rubber. Provinces have argued that replacing "poor quality" natural forest with rubber trees is not a significant change in ecotype, as it is just substituting one type of carbon-holding woody material with another (VietnamNet 2008). MARD, which had long opposed granting these cash crops "tree" status, surprisingly issued a decision in September 2008 (at almost precisely the same time that REDD+ began to take off as a policy) that finally agreed and declared rubber a multiple-use "tree" (Tô Xuân Phúc and Trần Hữu Nghị 2014).[25] Once MARD agreed that rubber was a "tree," localities felt free to expand rubber areas and call these "forests" or afforestation areas. Given that industrial rubber expansion in Southeast Asia has been implicated in major patterns of deforestation in recent years (Ziegler et al. 2009b), it is ironic that the ontological questions raised by REDD+ of what counts as a carbon-worthy tree may in fact further drive local deforestation.

This process of rethinking what trees and forests are in light of REDD+'s attention to carbon has also included a reassessment of the ways in which forests are classified in Vietnam. As noted in chapter 3, the 1991 Forest Protection and Development Law did not provide a definition of either forests or trees, and for many years, lands with no tree cover were classified as "forest land," over many local objections. This problem was finally addressed in 2009 when MARD issued Circular 34 that provided more precise definitions of forests, including a clear statement that lands without trees, including so-called bare hills, should no longer be classified as forests. Donors involved in REDD+ applauded this decision, noting that "Such definitions are helpful, as they prevent barren lands from being classified as 'forests' for the purpose of REDD+" (UN-REDD 2013, 8). Yet according to many people in the forestry sector, Circular 34's purpose has not been to clarify the definition of forest to improve REDD+ financing, but to allow powerful actors to legally claim lands for development that had once been considered off-limits. Many foresters and NGO workers in forest policy nicknamed this policy change the "Circular for Forest Destruction": what they meant by this was that by reclassifying lands like some vegetated bare hills, which may have poor stocking rates or small tree diameters but which were still previously considered as forest, and therefore could not be converted to other uses according to the old land identification system, now no longer count as "forests." These areas are now free to be transformed for other uses, including hydropower development or cash crops like rubber. While the 1991 "forest" classification was too loose, and captured many local areas for conversion to forests as I noted previously in chapter 4, the 2009 classification is likely now too tight, and is leading to the opposite problem of now-legal conversion of lands with some forest cover to agriculture.

Many previously classified forest areas are now rapidly becoming rubber plantations; in the Central Highlands, about 400,000 m^3 of timber has been harvested from so-called "degraded" forests occupying 700,000 hectares, which have then been converted to rubber (VietnamNet 2014). The financial incentive to convert is clear, due to high world rubber prices, and REDD+ projects cannot compete on favorable economic terms with rubber plantations at this time; carbon prices on the world market would have to be US$95 a ton to be equivalent to prices received by rubber farmers (Ogonowski and Enright 2013).[26] Lâm Đồng is one such location undergoing this land conversion frenzy; the province has signed contracts with investment partners for around 60,000 hectares of land slated for rubber development by 2020, which will require around 42,000 hectares of natural forest to be cut down. The documents on this conversion were passed to me under the table by several people in the provincial Forest Department, who believed strongly that rubber should not be considered a forest tree, and who expressed disbelief that MARD and the Prime Minister had changed their policy on this so suddenly. These whistleblowers also objected to the fact that most of the rubber plantation contracts that the province has already signed were industrial farms that belonged to state-owned or parastatal companies who had made powerful friends on the provincial political council, while only a few areas to be converted were to be reserved for smallholders. The fact that Lâm Đồng, the most active site for both PES and REDD+, has simultaneously pursued policies to protect "ecosystem services" at the same time that it has proposed replacing many areas of natural forests with rubber indicates that environmental rule often provides a cover of support for "green" policies, while at the same time hiding the actual implementation of anti-green actions.

But in a sign that not every tree is an equal target of environmental rule, one issue that has rarely been discussed among REDD+ projects is the carbon potential of swidden fields.[27] Like many other REDD+ interventions, Vietnam's projects have often targeted swidden agriculturalists as agents of forest destruction, and have recommended that REDD+ payments be used to halt these practices (UN-REDD 2013). However, studies elsewhere in Southeast Asia show that converting swidden agriculture to permanent agriculture or replacement by cash crops (including rubber or oil palm) likely results in net carbon losses (Bruun et al. 2009).[28] Yet discussions about the potential for fallowing swiddens to also sequester carbon have not been raised in Vietnam, where swidden continues to remain an enemy of foresters and forest projects, as it has for most of this century.

CONCLUSION

The saying "old wine in new bottles" is widespread in Vietnam (*bình mới, rượu cũ*), and is often applied to the numerous attempts that have occurred over the past thirty years since the beginning of *đổi mới* to reform the forest sector, from land allocation to reforestation funding to law enforcement to SFE reform. In many ways, PES and REDD+ may be old wine in as much as they have relied on the same sorts of problematizations and interventions as previous forest policy approaches. Despite the new label of "market-based" for these policies, PES and REDD+ have been largely justified by continued problematizations about forested uplands being necessary for downstream water supplies, and ethnic minorities as being ecosystem destroyers who do not understand the value of forest protection. But ultimately, PES and REDD+ as exemplars of environmental rule have been more about reclaiming new forms of financing for continued state management over remaining forests than they have been about new approaches to forest conservation that might rely on decentralized or financialized market transactions. Other social and economic justifications for these policies have included competition between state ministries over funding and power, and concerns about ensuring the continued supplies of water and electricity to wealthier downstream communities.

In terms of interventions to change conduct, these "market" approaches also look remarkably similar to previous state-centric eras. The PES projects that have paid households for watershed protection, funded via surcharges on water use by urban consumers and hydroelectric generators, have not been much different than older payment-for-protection policies noted in the previous chapter. PES and REDD+ programs continue to use state apparatuses, such as SFEs and forest departments, as managers of forest lands; state-trained forestry officials and "experts" to measure carbon and classify forests; and for households to serve as the main labor for forest conservation and replanting activities. There were no markets to be seen in any of the projects that actually operated on the ground, or in the policies that the government adopted; rather, strong state control over the parameters of policy continued, as exemplified in the nationwide pricing for PES water and energy fees.

This is not to say that the new policies have been mirror copies of old ones. While similar in many ways to old forms of environmental rule, one difference stands out: in these new "market" approaches the means to change conduct was determined first, while problematization and knowledge construction followed. That is, the intervention was determined at the outset: payments were the driver of policy, and only secondarily was concern given to what environmental services were or where carbon-rich areas could be found. Unlike other projects of environmental rule where the intervention was made to fit the knowledge and "problem" that had been produced, in these cases, the knowledge was filled in later after the intervention had been chosen. In reality, few PES projects conducted new research on actual environmental services or flows within ecosystems, or local valuations of the multiple values of forests. Unlike the often complicated constructions and circulations of knowledge that characterized previous eras of environmental rule, the projects examined here were based on simplistic external assessments about links between forest cover and water flow. Further, the focus on ecosystem services "users" as the drivers of policy has meant that the creation of knowledge has been rudimentarily condensed in both ecological and social realms: the ecological values of forests have been nearly always reduced to easily-measured water flow equations, and the social values of forests have been nearly always reduced to a simple quantitative number on the price that should be charged to a consumer.

In other ways, PES and REDD+ have not been much different than previous iterations of environmental rule. Problematization around deforestation in Vietnam has long focused on some scapegoats, while ignoring other drivers, and the fact that these projects have continued to target ethnic minorities indicates a continued concern about individual culpability for deforestation, while ignoring the state history of deforestation as practiced by SFEs and other government organs. In most pilots of PES and REDD+ in Vietnam, the subjects who must change their conduct are primarily upland communities, with little asked of other more politically powerful actors. While much attention has been paid to identifying water quality and flow as important ecosystem services provided by forest conservation in Vietnam, there has been little attention to actions that lowland or urban populations might be asked to make themselves to avoid water pollution or overuse.

Further, interventions have created new environmental subjectivities: upland households as "providers" of ecosystem services. But such conduct-changing has not been without resistance. When PES pilots attempted to develop tiered pricing so that different service providers were compensated according to K-coefficients for conditionality, local level actors perceived this as unequal and unjust. These outcomes suggest that externally-oriented valuation—such as the willingness to pay for water in downstream urban households in the PES projects, or the global price of carbon that is likely to influence payments in any REDD+ project—are not commensurate with local values of equality and labor compensation. This argues for a more complex accounting of power and social interactions within PES and REDD+, which to date have been mostly ignored by funders.

Yet by all accounts, PES continues to pick up steam in Vietnam, with a multitude of new donor projects and the attention and support of the state. These networks have been important in pushing forward new iterations of environmental rule: the enrollment of both donors and the state in enthusiastic support raises interesting questions about the process of translation, as global policy moves to local implementation, and indicates that boundary objects like the idea of ecosystem services or market-driven conservation can be many things to many people. As scholars of actor-network theory have long posited, circulations of ideas among networks of actors can produce new forms of materiality and objects, and we can see this process happening in Vietnam as REDD+'s focus on carbon and PES's focus on ecosystem services opened up an unexpected Pandora's box of questioning on what constitutes a "tree": an ecological definition or a carbon definition?

These new ideas about carbon and tree cover for hydrological improvement have allowed for a reassessment that coffee and rubber trees, since they too contain carbon and anchored soil, might be for the first time considered "forests." Did discussions over REDD+, which raised questions about what trees and forests would be considered valuable in the future, help facilitate an opening whereby these land re-classifications that benefited industrial interests could occur? Or did REDD+ participation simply provide a "green" cover for provinces and development agents that were intending all along to push for conversion of poorer quality forests to cash crop expansion, especially at a time of global commodity boom prices? The answer to these questions is not clear, but certainly the entire process lends credence to the idea that environmental rule facilitates certain justifications for policy actions, all the while obscuring other reasons. Lâm Đồng Province's ability to promote itself as a pioneer in both PES and REDD+ projects to preserve "green forests" and ecosystem services, all the while the local authorities were promoting forest destruction and conversion to other forms of development, certainly speaks to the power of environmental rule to facilitate negative outcomes, all in the name of "saving nature."

Conclusion

Environmental Rule in the Twenty-First Century

AS I COMPLETED THE REVISIONS TO THIS BOOK IN MARCH OF 2015, citizens of Hanoi took to the streets, singing songs, holding up signs and banners, and urging one another to engage in civil disobedience, all to save some trees. This outpouring was prompted by what the protestors considered an ill-informed and secret decision by the city's Department of Construction to replace 6,700 "old," "unsightly," or "dangerous" trees along the city's roadways with newer "modern" varieties that could withstand pollution and storms, and would not poke their roots through sidewalks and streets. But the middle-class residents who flocked to Twitter and Facebook to promote campaigns called *#6700cay* (6,700 trees) and "*6,700 người vì 6,700 cây xanh*" (6,700 people for 6,700 green trees) did not see the justifications for this tree removal policy as nearly so benign. Conspiracy theories circulated that the huge hundred-year-old trees were being cut and sold for valuable lumber to line the pockets of officials, or that the new trees to be planted, which were supposed to be native high-value *Magnolia fordiana*, were being swapped with a related low-value tree in the Magnolia family so that the contractor planting the trees could steal the difference.[1] Posters declaring "*Tôi đang khỏe mạnh, xin đừng giết tôi*" (I'm healthy, please don't kill me) began popping up anonomously on trees across the city. Protest marches took place in public parks and along major arteries, with monks, school children, musicians and pop stars, scientists and scholars, and everyday citizens joining together to demand answers and "transparency" (*minh bạch*) as to why the trees were being felled, and eventually the head of Hanoi's People's Committee called a halt to the cutting less than a week after it began.

Like other policies of environmental rule implemented in Vietnam, Hanoi authorities had used environmental reasoning to promote a preferred state outcome (different species of trees along city roads), leaving the true justifications for the inexplicable suddenness of the tree felling for citizens to try to figure out. The immediate distrust that this policy was about nature and growing conditions for trees, and the suspicion that it was more about corruption and financial incompetence, gives an indication that environmental rule no longer operates quite as well as it used to in Vietnam. Certainly the setting of modern Vietnam in 2015, with Internet access, global media attention, and a wealthier urban middle class, accounts for some of the reasons why environmental rule is changing. But even in these conditions, authorities still believed they had the proper "scientific" knowledge about species, the legal authority, and popular support to engage in tree replacement without need for citizen consultation or even information sharing before it happened. In other words, although the reactions to environmental rule may be evolving in the twenty-first century, the conditions that give rise to regimes of rule in the first place appear to have changed little.

As this book has argued, environmental rule is not a specific policy, and not a unitary philosophical or theoretical approach to nature, but rather an amalgam of practices that may shift in emphasis and outlook to assert environmental solutions for social problems, but which largely rely on similar organizing principles of problematization, classification, intervention, and subject formation. In the chapters preceding, I have relied on empirical and ethnographic evidence to help identify how environmental rule emerges and evolves, as the forms it has taken have changed over time. Using this historical-ethnographic approach has usefully revealed that different eras featured divergent techniques and technologies of rule as well as different authorities and actors, but all policies shared in common a focus on "problems" beyond those strictly confined to nature. Over time, environmental rule in Vietnam has been about such disparate goals as economic development and raising of revenue for the colonial state, about population resettlement and redistribution in an emerging postcolony, and about enrolling households as new economic managers of land in a new market economy under *đổi mới*. What different eras and approaches to environmental rule share in common is a more or less deliberate attempt to focus attention on the environmental aspects of policy, on "nature," when in fact it is the social aspects, or "society," that are the true targets and beneficiaries of rule.

Yet environmental rule is a tricky concept to translate into Vietnamese, despite this country being a remarkable case study of the diffuse forms that environmental rule can take. It is not an emic term, and even trying to directly translate "environment" (*môi trường*) and "rule" (*cai trị*) provides a less than satisfactory definition. To add to the difficulties, the idea of "the environment" or "nature" has long occupied an ambivalent position in studies of Vietnam's history, culture, and society; the two realms are rarely considered together. A commonly encountered sentiment regarding the role of "nature" in Vietnam can be summed up by a statement by well-known historian and literary scholar Neil Jamieson:

> Few Vietnamese valued the jungle, the mangrove swamp, or the forested hillside; they valued the bamboo hedge around the village, the banyan tree near the village gate, fruit trees in the garden and a shade tree in the yard, the village well, fishponds, a pier in the pond, the bunds around the rice fields, irrigation canals, dikes along the river. Remote and exotic flora and fauna, like jungle and swamp, lay outside of their systems of meaning, beyond culture. Such things were useless, even dangerous. (Jamieson 1991, 8)

The artificial divide between ideas of nature and culture has thus been alive and well in Vietnam, making it difficult for scholars to see how profoundly important environmental issues and policies were for the development of the nation's history (see Biggs 2012 for an exception). Yet the emergence of Vietnam as a nation was intimately tied up in the process of nature-making, and through the lens of environmental rule, important processes of state formation, contested citizenship, and shifting subjectivities can be seen. Far from being sites of fear and "beyond Vietnamese culture," forests have been essential to the birth and growth of the state: forests served as staging areas for the mobilization of workers moving from deltas to the hills to build a new socialist nation, for funding the treasury in grim post-war years, and for connecting Vietnam to new global networks of conservation and development in the open economic era.

This book has tried to bring these two concepts of nature and culture back together with a detailed examination of forests as not a natural ecological state, or even as simply a social construction, but primarily as an "effect" of environmental rule (Painter 2010). Within Vietnam, forests are an effect in multiple ways; they are an effect of the state, of local practice, and of social relations and networks. We can see forests-as-an-effect with regard to the state by looking at the fact that the current distribution of forests is largely a result of histories of where French forest reserves were located to remove timber; where SFEs were located for reasons of security, such as near ethnic minorities; where protected areas and watershed forests were established, and where biodiversity values were promoted by scientists and NGOs; and forests in the future are likely to be an effect of where and how carbon is valued. The species composition of forests is also a state-effect, as French and DRV logging removed most valuable hardwoods, and illegal logging driven by logging bans in the 1990s targeted the same, while state afforestation policy has focused on exotic fast growing trees. Forests-as-an-effect of local practice and social relations can also be traced; economically valuable products from trees have long circulated in local systems of exchange and use, and trees have been replanted on lands classified as "bare hills" through affective human labor. Forests-as-an-effect reminds us that we need to look at the processes by which forests are made, rather than the end results of these shifts and relations.

Conversely, we can also see people-as-an-effect of forests and forest policy: where people are settled and the types of livelihood activities they engage in is a function of how forests have been managed and classified over the past century. The millions of peasants who resettled from lowland deltas to upland forests during the 1960s and 1970s as part of state migration and employment plans have had their identities and subjectivities formed by these important events. For the additional millions of ethnic minorities who have long inhabited upland areas, social policies of sedentarization, biodiversity conservation and protected areas, and land allocation policy have all served to shape how individuals and communities are supposed to relate to the surrounding environment. Shifts in state policy have left households near forests often uncertain as to how to act, unclear as to what the state role is, and what their subjectivities of care for forests as households should be.

Treating forests-as-an-effect shows us that—in Vietnam and elsewhere—environmental rule is always contingent, and we need to make sense of the processes and pathways by which it occurs, rather than looking at before and after states, such as "bare hills" to forests, or deforestation to afforestation. The shifts in goals and patterns of effects imposed upon forests make it impossible to follow a single structural narrative or theory to understand forest change. Different waves of policies were each contingent moments that sometimes reproduced similar outcomes, but in other times did not, helping us see forests more clearly as sites of social enactments of various kinds. Environmental rule helps us work through these problems and proves invaluable in sorting through questions of cause and effect, reminding us that extended genealogies of processes helps us understand often inexplicable and contradictory outcomes, rather than taking moment in time snapshots.

THE "HOW" AND "WHY" OF ENVIRONMENTAL RULE

Treating environmental rule as an object of study fits well with long-standing concerns in political ecology as a field, as political ecologists have successfully argued that so-called "environmental" problems should be denaturalized and reformulated as social problems. In this way, understanding the "how" of environmental rule, namely problem identification, knowledge production, conduct shaping, and identity making, provides pathways to see processes of nature-society formation in action.

Problematizations under environmental rule have shifted over time, but a few key "problems" of nature and culture have emerged as persistent targets of rule: lands that were absent trees (namely, bare hills), and citizens who were absent state-approved cultural and agricultural practices (that is, ethnic minorities practicing swidden or peasants around government forest areas refusing state authority). Representations, rather than reality, have been key in identifying these problems, whether it was "deforestation" or "bare hills" or "swidden," which in turn have been fundamental in reshaping landscapes, and, oftentimes, in enacting dispossession as well. Close attention to environmental rule shows how shifting definitions and representations of forests may play a larger role than the actual policies themselves in shaping conduct. For example, the definitions of vast areas of the rural countryside as "wastelands" demonstrate the importance of paying attention to representations of forests, rather than their actual biological essences.

Conducting conduct has required manifested practices, including gazetting, posting regulations, policing boundaries, resettlement of populations, shifts in land tenure access and ownership, and checklists, contracts, and pledges of duties and rights with regard to resource measurement and management. During the French colonial era, the Forest Service was authorized to demarcate and define what forests should be reserved for scientific management through methodological felling, based on their experiences with Alpine forestry back in France. Later institutions, like SFEs under the DRV, or protected areas management boards in the *đổi mới* era, continued this strong role for the state in defining and reserving specific forest lands and determining what activities could be done on them, as was done, for example, through patrols by KL rangers. But not all forms of conduct-shaping have been state-led or coercive in Vietnam's forests; new attention to environmental services has encouraged local communities to engage in forest patrols in exchange for financial support from donors, and requests to citizens to monitor their own forest use in the name of "development" and "modernity" have also been made. Thus appeals to subjectivities have been a fundamental part of the story as well.

All of these processes have been assisted by networks of actors and objects that have circulated around environmental rule, as actor-network theory has pointed out. These networks have included state actors with myriad positionalities (colonial foresters trained in France, fervent socialists committed to building a new nation, forest rangers susceptible to corruption, or ministry officials looking for funding); international "experts" and scientists (French botanists, East German silviculturalists, Swedish development donors, British conservation biologists, and North American hydrologists); donors and financial supporters (Russian military aides, UN technocrats, academic economists, NGO workers engaged in poverty reduction, and international carbon market developers); and local citizens (forest product collectors, swidden agriculturalists, migrant SFE workers, illegal loggers, land claimants, forest patrollers, or providers of environmental services).

Yet, we should not treat either networks or actors as predetermined positions or categories; as actor-network theorists point out, they "have no wish to start with fully formed actors, identities, categories or dualisms; rather, they wish to trace the emergence of these as associations are formed, roles are defined, and divisions are established" (Murdoch 1997, 744). Networks of environmental rule have been fragile, tenuous, and ephemeral, often reconstituting in different forms at different times. Roles and associations have come into being and diverged again in different projects of rule: the network of colonial foresters who enrolled shipping companies, agricultural concessionaires, and Vietnamese mandarins in support of expanded French Reserved Forests in the first half of the twentieth century collapsed after World War II, to be replaced by a completely new network of nationalist politicians, inexperienced forest recruits, and experts from the Eastern Bloc working to build SFEs. Some actors in the network would not exist except for the actions of others: "illegal loggers" (*lâm tặc*) of the 1990s only emerged as actors thanks to the reclassification of certain forest activities as contrary to state law. And "ecosystem service providers" were created overnight with the signing of a state policy to pay households for forest protection dependent on where they were living within watersheds.

Networks have also included objects as actors: tree seedlings, land tenure certificates, maps of forest cover, and cash payments for environmental services. For example, tree seedlings in afforestation campaigns become endowed with normative capabilities: they helped forest-users become "good citizens" through planting and care of new saplings. By looking at objects in these networks, environmental rule reminds us that forests are not only epistemologically shape-shifting, depending on particular approaches to knowledge production, but may be ontologically shape-shifting as well. For example, many households living around Kẻ Gỗ contested the idea that forests even existed: they saw lands that were plantations, or bare hills, or pastures, or fields with trees, but not truly "forests." In another example of ontological shifts, new forms of value embodied in global REDD+ policies have brought forth an emergent commodity with no local price or tangibility. Thinking of trees as stocks of carbon, an immaterial concept to most, has required a new conception of what a tree or a forest is, leading developers of rubber plantations to take advantage of this new ontological space.

Projects of environmental rule were transformed by these interactions of actors within the network. For example, the shift from state-centric environmental rule to a more individualized and market-oriented approach after the turn of the twenty-first century was in some fashion a result of both counter-conduct and co-production of new forms of environmental valuation and subject formation that had happened under *đổi mới* previously. Households refused certain types of authority over forests, such as state KL rangers, while at the same time certain trees refused to grow as directed during afforestation operations. As a result, subsequent campaigns focused more on encouraging self-care and choice in participation by households, whether in tree planting or environmental service protection, and on opening up more expansive definitions of what constituted a tree or a forest.

In addition to paying attention to the "how" of environmental rule, we also need to identify the specifics of "why" environmental rule has been enacted. Forests were often shaped by practices and actions that were primarily aimed at other, larger goals; under the French, management of forest reserves was more about financing an underfunded imperial enterprise and controlling native populations unsettled by colonial rule, while under the DRV, forests were the sites in which industrial socialism that united ethnic minority peasants with logger-worker proletariats could be put into practice. Reforestation projects in the 1990s had goals of land allocation, protection of watersheds for downstream food security, and prolonging state employment in SFEs, rather than simply concerns about the expansion of forest cover alone. Environmental services policies in the 2000s emphasized new sources of domestic and international financing for the state, driven in part by bureaucratic power struggles between forest ministries and those in charge of industrial development.

These submerged histories and goals for rule raise the question of intentionality: is environmental rule a deliberate pretext to hide social goals under environmental practices, or is it more diffuse and less directed? These are questions that are similar to those raised about studies of resistance: to what degree does resistance require deliberate recognition of the fight against power, and to what degree does it arise spontaneously (Scott and Kerkvliet

1986; Ortner 1995)? The case of Vietnam at least shows that both types can coexist, and if we want to analyze the emergence of environmental rule in different situations, we should be cognizant of each.

As the previous chapters have noted, purposeful as well as organic drivers of environmental rule can arise over time. While some justifications of policies have been nakedly obvious, like officials who spoke to donors of biodiversity conservation as the reason for reforestation campaigns while funding species that actually benefited pulp and paper mills, other drivers of rule were not deliberate attempts to hide social goals, but rather were built on faulty knowledge or reasoning about nature: policy was often made on the basis of incorrect problematizations, incomplete knowledge, or deliberate ignorance of the knowledge that had been created. Policies to halt forest degradation usually classified only certain actions in forests as "deforestation," such as swidden agriculture, while ignoring others, like state logging, which led to the unsurprising failure to halt forest decline. In another example, the creation of knowledge regarding the value of forests for payments for environmental services was limited to easily measured services like water, while activities that degraded forests but which had no clear "users" that could be charged a set fee, like illegal logging, were ignored. In both these examples, it was a particular cultural or social viewpoint that directed action towards some "problems" of nature, and not others, whether or not such attention was deliberate.

In still other cases, there were actors that felt they were making policy aimed at environmental protection with no underlying hidden or cultural agendas; the biologists and other scholars who campaigned for national parks in the 1960s as an antidote to SFEs were genuinely concerned about the structural functions of forests that were being degraded by state mismanagement. And many conservation NGOs working to expand protections for endangered species in the 1990s relied on what they considered objective assessments of species rarity and threats to suggest protected areas expansion, even though these were largely ignored by the state in favor of sites of former SFE logging. In other cases, these conservation actors had but a partial focus and understanding of a problem, such as concern for the spatial delineation of protected areas, but without a simultaneous concern for the need to involve local people to run these areas. Thus it was not that no one had conservationist sentiments; it is that these were not the dominant reasons or justifications for environmental rule, or these advocates often were not able to enroll others in their concerns, or they were coopted for policies with other less benevolent goals, as was the case with conservation NGOs who ended up supporting reforestation campaigns that actually diminished Vietnam's biodiversity.

Understanding these hows and whys of environmental rule has allowed us to better see how landscapes came to be "thought of" and "calculated for" and "subjected to" numerous initiatives, some of which initially seem quite contradictory (Li, 2007b; Miller and Rose, 2008). In the remaining sections of this conclusion, I explore more in-depth the ways in which knowledge, actors, and power came to shape forest policy in the specific context of Vietnam, in the hope that this attention provides a useful blueprint for other studies of environmental rule in other countries and other situations.

KNOWLEDGE REVISITED: HOW IS KNOWLEDGE DEPLOYED UNDER ENVIRONMENTAL RULE, AND WHY?

Of all the practices that make up environmental rule, knowledge formation has been fundamental to the creation and management of forests in Vietnam. Forest definitions and classifications have shifted over time, and it has always been a political act of rule, rather than a "scientific" one, to define certain types of forests and declare them as worthy of attention, while others were not. Valuing some types of forest products, such as timber from hardwoods, while other forest products and services, such as wave protection afforded by mangroves or fuelwood collected by women, were ignored, was similarly a political choice. It is thus important to examine in any case of environmental rule how divergent understandings of and knowledge about nature and the environment evolve, and how policy responses to this knowledge also emerge.

The generic concept of "forests" served as a boundary object par excellence throughout different eras in Vietnam; the fact that forests easily become a slot into which concerns over revenue, development, cultural practices, state control, biodiversity, or climate, could all be put points out how important certain biophysical properties are in social relations and imaginings (Prudham 2003). Ideas of "forests" were so porous that wildly divergent organizations could be brought together in networks to support activities in these areas that might have made no sense in any other context. For many of these networks and associations to work, however, different types of knowledges had to be generated at different times and for different audiences.

For the French, a true forest was a place that had been inspected by the Forest Service and where a "scientific" coupe cutting plan was in place. The classification schemes deployed by the French Forest Service in Indochina nearly always used both "scientific" assessments and economic valuation as a basis for differentiating between forest types, such as "rich" or "poor" forests, which referred to the extractive value of timber and density of stands. For the DRV, a forest was essentially any area that could be brought into timber production through creation of an SFE; unlike in the previous French era, "science" was rarely deployed for forest development, but rather "expertise" was provided by forest engineers and economists who assessed how much wood was needed by the state, not on what extraction was appropriate for forest structure. For environmentalists and international conservationists in the 1990s, a forest was a place with high biodiversity value that hopefully could be delineated as a special-use protected area, although the state still used inconsistent definitions of what "forests" were in reclassifying land as special use, protection, or production forests. Further, the government definition of "forest land" as "places the government has classified as forest" clashed with definitions of forest held by local people that distinguished between visible ecological states (such as land with few trees or land with many trees). In the 2000s, for state forest officials, a forest was a place where carbon or other environmental services were stored and international funding could be obtained, particularly in watersheds. These subjective and shifting definitions of forest remind us to always raise questions about treating forests as an object with a shared semantic meaning, and to pay attention to the ways in which co-production results in transformations of definitions and concepts (Forsyth 2011).

One important component of knowledge production and problem definition for forests under environmental rule have been tools and technologies of calculation, and in particular practices of mapping, which enables simplification, legibility, and rationalization of space. Forests were visualized as sites of timber extraction in the early 1900s to 1990s, sites of biodiversity reservoirs from the 1990s to 2000s, and sites of potential for afforestation where bare hills were demarcated from the 1990s onward. Forests in the early twenty-first century in Vietnam became reenacted anew by practices of valuation and mapping as areas that provided watershed services and carbon sequestration became newly "seen" and embodied. Threats to forests were visualized as well. Some threats came from material inscriptions on landscapes: a burnt plot of land, a farmer in ethnic garb in a swidden field, an "illegal logger" removing timber from state lands. Some threats were invisible and could not be visualized, such as the threat of a warming planet, but were nonetheless important in imagining certain types of forests as endangered.

Just as certain types of knowledge were valorized, particularly that generated by the state and "experts," other knowledges were either ignored, or silenced. Local rules for common property practices of forest management were of little interest to the French Forest Service, who for many years declared that only the state should be allowed to designate and run Reserved Forests. In the DRV era, the knowledge held by ethnic minorities about the forests of the uplands, to which millions of people were to be moved, was similarly of no use, and was to be replaced with knowledge generated in the service of industrial socialism. Later, after *đổi mới*, while there was plenty of attention to the seemingly widespread nature of bare hills, there were no government studies on what vegetation might already exist in these areas or what the value of these "wastelands" might be to the poor or to women. And given the considerable amount of attention paid to eliminating swidden agriculture across the twentieth century under different projects of environmental rule, none of them attempted to produce knowledge that actually might understand the environmental adaptiveness of this agricultural practice, despite strong existing evidence that swidden is a far more environmentally sound practice than many other alternative land uses (Ziegler et al. 2009a).

Knowledge circulation through networks has been key in both establishing authority, as well as consolidating standard tropes and discourses, as distribution and exchange of information occurred throughout Southeast Asia and globally (Vandergeest and Peluso, 2006a). French forest codes moved from Paris to Hanoi to other French colonies with ease. Networks of forest knowledge between the DRV and East German and Soviet forest experts in the 1960s influenced the places where SFEs were located and the activities that these SFEs engaged in. Small networks of Vietnamese scientists were able to establish Vietnam's first protected areas, and later joined with foreign donors and conservation NGOs to extend biodiversity protection to other forests. Global networks of scientists, donors and NGOs concerned about climate change similarly introduced new forms of carbon and ecosystem services value into Vietnamese forests in the 2000s. In each case, key figures enrolled others in the network through *interessement* and encouraged the circulation of standard narratives of knowledge.

As other studies have pointed out, however, circulation alone is not a sufficient condition for knowledge dissemination: different knowledge claims count more than others (Mathews 2011). Not all actors or networks are equal, which raises questions of whose claims count and whose are legitimized by power. Multiple knowledges were often at work, requiring relations to sustain some knowledges as superior to others, as so-called "expertise" over forests in Vietnam has shifted over time. What constitutes expertise under environmental rule has never been consistent, but expertise that made universalizing claims, or which was generated in formal institutional settings, was often privileged over others. Knowledge of forests that was local and specific was often targeted for elimination, to be replaced with more "scientific" claims of value, such as global carbon stocks.

Ultimately, the knowledge that emerged from different eras of environmental rule, and the forms of circulation that these knowledges took, have always impacted the on-the-ground social and development interventions and outcomes that resulted. Specific types of knowledge directly fed into practices of social control, including anti-swidden campaigns, the development and securitization of upland areas though resettlement and migration, or donor assistance for forestry and climate adaptation. Processes of land alienation and dispossession were made possible by the ways in which deforestation and bare hills were problematized and knowledge about these areas circulated. Knowledge that created spatial reclassification of forests through new technologies often simultaneously resulted in new social classifications of forest-using peoples as well, as people were sorted into categories of "illegal loggers," "weakly settled swiddeners," or "environmental service providers."

ACTORS REVISITED: HOW HAVE ACTORS PLAYED ROLES IN ENVIRONMENTAL RULE, AND WHY?

Actors in forest policy in Vietnam have been highly diverse, from Communist Party officials to international donors and emergent NGOs in civil society, and, complementing this, subjects of forest policy have shifted as well over time. Subjectivities among these actors have strongly been shaped by networked relations, as well as the fact that people's own experiences and identities often can alter in relation to environmental change (Escobar 1999; Swyngedouw 2004). Some subjectivities arose organically, while others were objects of manipulation from projects under environmental rule. For example, nationalism has often been invoked in the name of nature making (Cederlöf and Sivaramakrishnan 2006), and, in Vietnam, appeals to state building were one component of rule. Some appeals were obvious and insistent, as during the early migration campaigns to get millions of lowland peasants to build socialist paradises in the upland jungles, while other appeals to nationalism were more muted. For example, the tree planting projects that often conjured the name and image of Ho Chi Minh continue to this day, reminding households year after year of the birth of the postcolonial nation, symbolized in the emergent young tree seedlings planted during the Lunar New Year's festivals.

Similarly, becoming subjects under environmental rule has required identity reconfiguration in different ways depending on forest relations and practice. Forest subjects were remade by different technologies of conduct, deployed at different times, and forest environments themselves were used to create these new subjects. Through migration and resettlement of lowland dwellers aligned with the developmentalist goals of the DRV, practices that enabled certain subjectivities show us that forests were often made by labor, and in turn labor and identity were often remade by forest work. New SFEs were manned by lowland recruits, the majority of them women, who arrived in the highlands and began new lives as forestry workers in what planners hoped would be a new socialist paradise in the hills, while sedentarization projects moved ethnic minorities into concentrated settlement sites where they could be "improved," despite the fact that little environmental investment could be detected in these zones. These people who practiced swidden agriculture, community land tenure or living in longhouses had cultural identities that were unacceptable, and needed to be replaced with those of the "new socialist man."

Forest rangers too were important actors affected by multiple subjectivities; they maneuvered between the feeling that they had little support from the state, consoling themselves with heroic songs, all the while serving as the objects of ire from local peoples who believed that rangers were the physical manifestations of unfair laws. Temptations from corruption and policies that encouraged illegal logging in the wake of logging bans further complicated rangers' positions. *Lâm tặc*, who were reviled in national media and blamed for much of the deforestation of the 1990s, saw themselves as simple peasants establishing claims on resources that richer state actors were unfairly accessing. The many cases of official corruption in the forest sector that were uncovered and which implicated officials from many government offices served as further evidence for the moral claims of the *lâm tặc*.

More recent technologies of conduct under environmental rule have focused less on coercive practices to change identity, and more on education work to instruct local peoples in the values of forests, whether as watershed protection zones, homes for endangered species, or stocks of carbon, and citizens have been asked to reflect on biodiversity conservation and carbon sequestration as modernist goals in a globalized world and remake their conduct appropriately. Gendered subjectivities have also been remade, sometimes inadvertently, as SFEs recruited mostly women, many of whom saw their work for the state at SFEs as akin to a marital contract, while later tree planting campaigns displaced women's resource use and put more forest land under the control of men. These subjectivities were not set in stone, however, as individuals and households often remade their relations with each other and with forests depending on circumstances at the time: today's illegal logger might be tomorrow's tree planter.

Actors in the networks of environmental rule that circulated throughout Vietnam's history have not been confined to human subjects. Indeed, as the histories outlined here have shown, it is also the material things in the network that also act, and we need to understand "the capacity that nonhuman nature and technologies have to shape the direction and character of socioecological change" (Birkenholtz 2009, 120). Forests themselves have influenced how people have related to them and tried to manage them: trees can become true actors in myriad ways.[2] For example, French foresters who wanted to convert Indochinese forest to clear-felled coupes found themselves stymied by high diversity of species in forests. Farmers who were encouraged to plant trees in the reforestation campaigns found themselves unable to force trees to grow in many areas without significant labor, capital, and water. In many ways, the ecological materiality of forests has required certain types of management practices and thus foreshadowed some of the forms of environmental rule that were applied. For example, Karen Bakker has discussed how the fluidity and bulk that characterize the physical properties of water have shaped policies for water management, namely leading to oversized roles for the state to play in construction of infrastructure (Bakker 2005). Yet unlike water, forests do not necessarily require extensive state infrastructure, although that can certainly be imposed (Demeritt 2001), particularly in large-scale monocrop plantations or extensive watersheds for ecosystem services delivery. Rather, forests lend themselves both to small-scale household management as well as top-down state management, leading to a diversity of interests that competed with one another, resulting in complex and shifting management arrangements over time.

Other objects too exerted influence on human management and policy; for example, the physical pieces of paper that French foresters tried to post on village walls that indicated their legal authority over forest lands became instead an object of contestation, as locals could protest that the paper did not arrive or could not be read. The maps that circulated, whether an imagined map of 1943 forest cover, or the later maps of watershed areas, also became actors in networks inasmuch as they caused other actors to take positions or make assumptions in certain ways. Tree seedlings as objects established claims to land for households during the decollectivization era, and allowed wealthier households to appear to be supporting state environmental plans, rather than selfishly claim-jumping. The payments that were brought to households for watershed protection became an object that locals could rally against in a fight for more equitable sharing of benefits from the new PES law. All of these physical objects were important actors in understanding why certain pathways of environmental rule emerged, and how.

POWER REVISITED: HOW HAS POWER BEEN MOBILIZED IN ENVIRONMENTAL RULE, AND WHY?

Political ecologists have long emphasized the need to pay attention to forms of power, and a great deal of focus in this field has located power within the state, particularly within the rights of state to define, frame and understand resources within national territories (Peluso and Vandergeest 2001; Sivaramakrishnan 1999). Such forms of power were often exercised in processes of territorialization over land, people and resources, and in disciplinary forms of control and punishment for violations of the state's right to claim forests, delineate protected areas, or allocate resource use rights. Environmental rule has also been associated with these sorts of interventions in disciplinary form: boundaries for reserved forests set up by French colonists with no local participation, and enforced by military and rangers; or the nationalization of all land over 25 percent slope under the later socialist state, managed by state forms of control and power in the guise of SFEs. Even later interventions that tried to emphasize concepts of "scientific" reclassification and reordering of forests that happened under the more open *đổi mới* era relied on forest rangers to patrol protected areas, with varying degrees of success.

But disciplinary power that has been long associated with the state is less prominent than it once was, particularly when it comes to environmental rule. Instead, newer hybrid, decentralized, and neoliberal forms of intervention have emerged, and understanding these shifts has required rethinking of different and novel forms of power. For governmentality scholars, this new concept of power is enabled through the ways in which "truth" about problems is produced, usually provided by those with so-called expertise. Bruno Latour and other ANT theorists, in contrast, have reflected on power as the ability to enroll and mobilize other people into networks to achieve goals (Latour 1987; Miller and Rose 2008). Both approaches remind us that power is often about control of ideas as much as it is about control of material goods, whether land or people. In the case of Vietnam, during phases of problematization regarding deforestation, certain actors and actions were targeted (namely, ethnic minorities practicing swidden) while other issues were overlooked, such as the inappropriate organization of SFEs and their role in deforestation, or the incentives for deforestation caused by other forms of policy, as in shrimp aquaculture or coffee cultivation. These selective interventions are a clear form of power.

The intersections of culture with power have been important in environmental rule as well, especially since many projects named as "environmental" in nature were actually aimed at changing cultural identities and practices. These identities related to the environment that needed changing included the consistent attacks on swidden that were also thinly veiled attacks on ethnic minorities' rituals, housing, or other cultural practices. But some cultural practices were not targets of environmental rule nearly as consistently; for example, the habits of urban Kinh to use luxury hardwoods and investments in housing as standards of middle-classness were never pointed to as drivers of the illegal logging that flourished in the 1990s, although every indication was that consumer demand played an enormous role in timber scarcity.

Similarly, resistance to environmental rule can be seen as an indicator of power as well. Even in Vietnam, where protest and political activism is muted due to the nature of the one-party state, citizens objected to impositions of environmental rule in numerous ways: environmental rule can never silence, even in a politically closed society. Despite many state attempts to remake new forest subjects, the objects of these campaigns often defined for themselves their own rules for forest use, including resisting state boundaries for protected areas, or the authority of new forest rangers in enforcing state laws. Resistance aimed at French forest restrictions and land alienation led to a political uprising in the 1930s that forever changed Vietnam. Resistance and foot-dragging eventually brought down SFEs as workers looked out for themselves and their families, using private household gardens and preventing regenerating activities. Yet in other cases where we might have expected resistance, such as the alienation of village common lands under reforestation projects, little action occurred, indicating that resistance cannot always be a predictable act.

One clear outcome that can also be seen is that protests against certain interventions in the name of environmental rule can also adopt the language and strategies of these projects as well. That is to say, protests that were really against social policies or injustices could be made into protests about the environment, in a convenient hijacking of justifications for rule. Indeed, it may be precisely because the nature of political activities is strongly curtailed in Vietnam that protests could be cast in "environmental" terms, presenting these as more apolitical than other types of protests. The protests against tree cutting in Hanoi in 2015 were explained in several media reports as precisely thus: citizens were actually angry at a number of things having to do with urban planning, lack of voice, lack of democracy, ineffective governance, or rising land prices driven by corruption, but those things could not be protested against directly. Trees, on the other hand, were seen as "apolitical" (BBC, 2015). Thus environmental rule and the use of alternative, hidden justifications for interventions can cut both ways: as tools for oppression, but also for liberation.

CONTINGENCY AND CHANGE IN ENVIRONMENTAL RULE

Careful examination of environmental rule upends many of the traditional narratives and assumptions about forests in Vietnam, from the idea that the Vietnam war was the largest destructor of forests, to concerns that deforestation in the uplands has been driven by Malthusian population pressures, to the optimism that reforestation is always environmentally beneficial. There is a much more contingent and nuanced story to be told, one that takes into account the importance of knowledge-production, the multiplicity of actors, and the intersecting forms of power that come together in not-always-predicable ways. By framing the understanding and use of forests in Vietnam through the lens of environmental rule, this book has pointed out the contingent and often unpredictable nature of change. Policy responses to problematization in the forest sector, which at the global level have most commonly focused on deforestation, often acknowledge only single or linear drivers of land cover change, such as export of timber, or shifting cultivation, or population growth, as these variables can be measured more easily. This approach, however, does not acknowledge the multiplicity of causation and the shifting temporal dimensions of environmental change that do not lend themselves as easily to fixed quantitative analysis.

Simplistic responses to simplistic problems in turn create feedbacks and drivers of still more land use change, as was the case with the logging bans in the early 1990s, which propelled illegal logging to new levels and brought new actors into the label of "forest destroyers." Proximate actors who might be most visible and legible to states—such as illegal loggers, agricultural migrants, or shifting cultivators—were often made scapegoats for these processes and were targeted to change their conduct, rather than developing policies that took a more comprehensive look at the underlying causes for these actors' practices. Such complexities in Vietnam might include the long histories of state control of forest lands and the application of state-determined definitions of what constitutes a forest; local distrust of state forest institutions as policies have shifted and been applied unevenly to different peoples; and competition over land and resource rights between and among communities and even within households between men and women.

There are great difficulties in overgeneralizing about environmental policy in one single country, let alone extrapolating to many others. Acknowledging contingency makes it difficult to predict how forest policies might result in particular outcomes in different places, especially if effects of these policies are unintended. As Tania Li has pointed out, many development (and forestry) projects are well-meaning, but interventions designed for one purpose often had effects that were contingent, diverse, and unpredictable (Li 2007). This can certainly be seen in the case of Vietnam. For example, there were moments were things could have gone very differently; this was not a trajectory that was preordained. For example, there was a brief moment in the mid to late 1950s when the anti-imperial project of the DRV tried to take seriously the idea that local livelihoods should be the driving motivation behind forest management, which could be decentralized and defined by local peoples, as I discussed briefly in chapter 2. The conciliatory tone towards swidden and local management of the late 1950s could have led to decentralized forest management, and the recognition that swidden was a necessary use of land for upland farmers in this period indicates a real attempt to reconfigure forest relations, but this moment was fleeting, only to be replaced by centralized forest bureaucracies that ended up reproducing much of the colonization of forests that the French had begun. There is no way to know what the counterfactual might have been. If the decentralized policy of forest management in the late 1950s had been continued, which promised localities the strongest role over forest management, would these communities have been invested in more sustainable forest management? If logging bans had occurred in 1973, rather than 1993, would that have reduced the need for illegal log imports from other countries? If forest land tenure in the 1990s had been given out in equal shares, as agricultural land had, would the growing gap between rich and poor in forest land holdings have been avoided?

Ultimately, such questions are unanswerable. What we can do instead is trace out the processes and networks by which environmental rule actually occurred, by paying attention to the ways knowledge is produced; the ways actors are empowered or disempowered; and the ways in which material landscapes and objects are incorporated into policy and practice all serve as key nodes for understanding trajectories of forest change. These pathways are not linear and as such they are often not predictable, but they are understandable and explainable within the framework of environmental rule.

Notes

Introduction

1 In Vietnamese, "Rừng là vàng, nếu mình biết bảo vệ, xây dựng thì rừng rất quý."

2 "*Rừng vàng biển bạc*," or "gold forests, silver seas," had long been a shorthand phrase to indicate the natural riches of the territory of Vietnam, was particularly used in nationalistic senses, and predates Ho's use here.

3 The entire speech was reprinted in *Báo Nhân Dân*, number 3453, 11-9-1963, located at http://www.na.gov.vn/sach_qh/chinhsachpl/phan3/p3_40.html [last accessed May 8, 2013].

4 The concept of co-production comes primarily from science and technology studies (STS) where scholars have argued that "co-production is shorthand for the proposition that the ways in which we know and represent the world (both nature and society) are inseparable from the ways in which we choose to live in it" (Jasanoff 2004, 2). Therefore understandings of natural phenomenon are always social, but are influenced by the material properties of the world as well (Latour 1993).

5 While Foucault was interested in the formation of human subjects generally, he focused particularly on the birth of the social sciences (focused on topics such as psychiatry, criminality, and sexuality) and did not have much to say about nature or ecology; it has been left up to others to apply his work on governmentality and on sovereign forms of power to the realm of the environment (for examples, see Darier 1999; Oels 2005; Agrawal 2005; Li 2007). Some of these works have focused on the idea of "green governmentality" (Rutherford 2007) or "eco-governmentality" (Goldman 2001), linking environmental subject formation to neoliberal projects of governance, which differs considerably from what I am suggesting constitutes environmental rule.

6 Nikolas Rose has asserted that the starting point for governmentality studies should be "asking what authorities of various sorts wanted to happen, in relation to problems defined how, in pursuit of what objectives, through what strategies and techniques" (Rose 1999, 20). Anthropologists have found this approach too authority-centric, and insufficiently attentive to what actually happens when idealized regimes of practice meet messy reality, such as the lived experiences of subjects who interact with authorities, and who often resist, alter, or thwart the strategies and techniques aimed at their conduct; in other words, what we might call governmentality-on-the-ground (Cepek 2011). Instead, these scholars have argued in favor of study of the untidy contingency of life, the "witches brew" of practices, in Tania Li's term (Li 2007), or the "micropolitics" of cultural struggles, in Donald Moore's formulation (Moore 2000).

7 Visibility is a tool "to picture who and what is to be governed, how relations of authority and obedience are constituted in space, how different locales and agents are to be connected with one another, what problems are to be solved and what objectives are to be sought" (Dean 1999, 30). As James Scott has noted, the ability of states to make people and landscapes more "legible" is a process that links such disparate undertakings as the use of standardized measurements, assigning of street names and patronyms, and government censuses and maps, all of which can be considered technologies of rule (Scott 1998).

8 In James Scott's formulation, these processes are state "simplifications" as states seek to order some problems solvable while ignoring others (Scott 1998). In James Ferguson's work on the problems of development in Lesotho, this process is termed "anti-politics," as political factors are wiped away in favor of narrow technical discussions (Ferguson 1994). Li relates a similar process in Indonesia, which she terms "rendering technical" (Li 2007).

9 As a point of comparison, James Fairhead and Melissa Leach's classic work on landscape change in West Africa has noted how narratives about deforestation are highly subjective (Fairhead and Leach 1995; 1996; 1998). Identifying processes of change as "deforestation," rather than the afforestation that the authors and their interlocutors saw, justified "repressive policies to reorientate what has been seen as destructive land management," including criminalization of many land practices (Fairhead and Leach 1996, 4).

10 There are many critiques of these orthodoxies, such as narratives about deforestation in the Himalayas; desiccation in north Africa and the Middle East; and global discourses about underdevelopment and poverty (Davis 2007; Escobar 1995; Forsyth 1996; Roe 1991).

11 Although in some of his 1978–79 lectures Foucault did discuss the idea of counter-conduct as a mechanism by which subjects became imbricated in the governmentality project (Foucault 2007), some have accused Foucault of glossing over the problem of resistance (Bevir 2010).

12 Foucault does allude to the importance of physical objects and events when giving a metaphor of government: "The fact that government concerns things understood in this way, this imbrication of men and things, is I believe readily confirmed by the metaphor which is inevitably invoked in these treatises on government, namely that of the ship. What does it mean to govern a ship? It means clearly to take charge of the sailors, but also of the boat and its cargo; to take care of a ship means also to reckon with winds, rocks, and storms; and it consists in that activity of establishing a relation between the sailors who are to be taken care of and the ship which is to be taken care of, and the cargo which is to be brought safely to port, and all those eventualities like wind, rocks, storms and so on; it is what characterizes the government of a ship" (Foucault et al. 1991, 93-4). Yet despite this promise, Foucault's thoughts on the ways material objects, events, and the environment might be the basis for rule need to be inferred more than seen in his writings.

13 The idea that non-human natures can be active constituents of a network is most associated with the work of Sarah Whatmore, who has argued in particular for animals as political actors (Whatmore 2002). Actants need not be living creatures, however, as

David Biggs has noted on his work on technologies of water management in Vietnam's Mekong Delta, "While some may wish to debate whether abandoned feasibility studies, broken-down equipment, and landscape features such as canals and airstrips have agency, when such objects are linked with influential humans, they may exert powerful influences on development work" (Biggs 2008, 622).

14 STS scholars have identified boundary objects as those "which are both plastic enough to adapt to local needs and the constraints of the several parties employing them, yet robust enough to maintain a common identity across sites. They are weakly structured in common use, and become strongly structured in individual site use" (Star and Griesemer 1989, 394). As Star and Griesemer have shown in their work, boundary objects serve as ways of problematizing and producing knowledge that are "simultaneously concrete and abstract, specific and general, conventionalized and customized" (Star and Griesemer 1989, 409). One important aspect of boundary objects within networks is that they serve to join disparate actors to work together in the absence of a formal consensus on ideas and goals (Star 2010, 601).

Chapter 1. Forests for Profit or Posterity?

1 Particularly influential in justifying these new actions were Antoine Fabre's *Essai sur la théorie des torrens et des rivières* from 1797 and Alexandre Surell's *Étude sur les torrents des Hautes-Alpes* from 1841. Clarence Glacken has argued in his magisterial review of Western attitudes toward nature that it was Fabre who served as the greatest influence on later attitudes towards ecological management and the links with natural disasters (Ford 2004; Glacken 1976). Surrell's later work applied Fabre's ideas to the specific case of the Alps, and "seduced foresters with an elegantly simple equation: 'Everywhere that one finds new torrents, there are no more forests, and everywhere that the ground is deforested, torrents have formed'" (Whited 2000a, 264). Despite the fact that Surrell used limited empirical studies of erosivity for his conclusions, he prescribed massive reforestation for all mountainous areas of France (Whited 2000b).

2 One of the earliest botanical surveys was compiled in the *Flora Cochinchinensis* by Portuguese botanist Joao de Loureiro in 1791, which names some seven hundred new species of plants (Merrill 1933); most of de Loureiro's collecting took place around Huế city. However, the *Flora* provides few details on the extent of forest cover during the mid-eighteenth century when the Nguyễn lords who ruled southern Vietnam employed De Loureiro.

3 International traders, often arriving on junks from Canton, regularly called at the port of Faifoo (now Hội An) in search of forest trade goods, making nearby Quảng Nam province in central Vietnam known as "the world's most fertile land," in the words of well-known court chronicler Lê Quý Đôn (1977 [1776], 337). In his *Miscellaneous Chronicles of the Pacified Frontiers* (*Phủ biên tạp lục*), Đôn recorded many of the products exported in 1776, including "black pepper, areca palm products, nutmeg, sapan wood, ebony, tropical cypresswood, cardamom seeds, rhinoceros horns, birds nests, deer tendons, fish fins, dried shrimps and prawns, 'fragrance snails' . . . tortoise shells, elephant tusks, crystal sugar, white sugar, iron, zinc powder, sea slugs, and hundreds of kinds of Vietnamese medicines" (Woodside 1995, 165–66).

4 According to Phan Huy Chú, the author of *Lịch triều hiến chương loại chí*, writing in 1821 for the emperor, taxes were levied on people and agricultural land, but not forest land, and only some forest products going to market would be taxed (Werner et al. 2012, 274).

5 In the comprehensive *Đại Nam nhất nhống chí*, a geographical record compiled in the late nineteenth century in which individual provinces' resources and infrastructure—such as schools, roads, markets, temples, minerals, and land types—were laid out in great detail, forest resources are absent from assessment and measurement. For example, in the volume published in 1882 outlining Hà Tĩnh province's resources, the emperor's assessors make no mention of forest area, despite the fact that this province would later be one of the largest logging areas under the French; the only reference to forests is a brief mention that *trầm hương* (*Aquilaria malaccensis*) collectors had to pay 1 kg of the product each year as tribute (Cao Xuân Đức et al. 1965 [1882]).

6 Arrêté réglementant le service forestier en Cochinchine, June 23, 1894.

7 While each colony/protectorate had its own legal system, forest laws were broadly comparable between each, with some local variations, such as the particular species that might be listed for special taxation or regulation. It was not until March 1930 that the governor-general ordered a forest law common to all five divisions of the Union (Madec 1997).

8 The 1827 Forest Code was similarly adopted wholesale for the colony of Algeria, despite the dramatic differences in climate and forest type between France and North Africa (Davis 2004).

9 For example, the new Indochinese regulations specified that in cases in which individuals who had violated forest laws, such as burning a pasture, could not be identified, entire villages would be punished, and, in this way, it was hoped that collective responsibility would prevail among citizens. These collective punishments were used throughout the French colonies after their initial use in Cochinchina (Baillaud 1914).

10 For a brief period of time in the early 1930s, a Graduate School of Agriculture and Forestry of Indochina (*École Supérieure d'Agriculture et de Sylviculture de l'Indochine)* operated in Hanoi and offered a three-year forestry diploma, but only seventy-five students graduated from the program before it was closed due to financial difficulties (Brown 1957).

11 It was not until the 1930s that Vietnamese were allowed to head divisions (Fangeaux 1931).

12 Maurice Mangin, who served as head of the Indochinese Forest Service from 1924 until the 1930s, argued that one reason for the slow and fragmented development of forest management was a move in 1913 towards decentralization, whereby the Forest Services reported to the Résident Supérieur of each colony rather than to Hanoi and the central government. This situation was reversed in 1924, and the Inspectorate in Hanoi once again regained centralized control, leading to the development of a single forest law for all five colonies in 1930 (Mangin 1933).

13 One official exclaimed that there might be one teak tree every three hectares in Indochina if foresters were lucky (Thomas 1998), whereas stand density in Indonesia for intensely managed natural teak or sal in India might be well over one hundred trees per hectare.

14 *Rẫy* is a word of Thai origin that has been adapted into Vietnamese as both a noun (a *rẫy* can refer to a swidden field) and verb (to '*phát rẫy*,' or make a swidden field). French officials noted that *rẫy* was not confined to Indochina; the practice was referred to as "*tavy*" in Madagascar, and some even noted that the French practice known as "*sartage*" in the Ardennes, whereby forest was cut and burned to plant rye and oats, could be considered a form of swidden (Mangin 1933).

15 Arrêté du 19 Mars 1907 concernant la constitutions de réservé forestier.

16 In one case of a reserve in Phan Rang province in Annam, for example, the official record of the local commission's meeting read as follows: "To be attending the meeting: the Résident Supérior of Phan Rang province, local administrators of two affected villages, the forest division and forest guard station personnel. Meeting: to set up 16,000 ha forest reserve. The representative villages to be affected by the reserve were not present, although having been duly cautioned, the committee considered their absence as a tacit consent on their part." NAV4 RSA 3353: Réserves forestieres Nho Quan, Calap (Thanh Hoá), Phan Thiết, Phan Rang.

17 It was only later that officials recognized these measures might be too harsh, and engaged in discussions about the possibility of reserving communal forests for exclusive use of native populations. By the late 1920s, new steps were being envisioned, given concern that the "meddlesome interference" of the Forest Service in communal land "would anger the population," according to a letter from the office of the Résident Supérieur (RST) of Tonkin. The RST noted that a revision of the Forest Law that recognized communal land title might be needed, and as long as villages had clear property rights to the land (such as proof of taxes), customary rights to cut timber could be acknowledged. NAV1 RST 78.394: Note sur la lettre No 4942 du 18 Dec 1925 sur les bacs et forêts communales.

18 Arrêté du 5 Sept 1905 réglementant les coupes en domaine forestier réservé.

19 K. Sivaramakrishnan has called these working plans "instruments of remote control." As he notes, "Working plans came to symbolize this confidence of the scientific forester; but visualizing a terrain where science could plan unimpeded the manipulation of the forest, compelled a more complete enumeration and disposal of local rights that might obscure the vision" (Sivaramakrishnan 2000, 65).

20 NAV4, RSA/HC 1086: Rapport de gestion Année 1919 Service Forestier.

21 New technologies emerging at this time were crucial for visualization and mapping; as officials noted, "aviation, by direct observation or photography, has emerged as the best and most economical way to general exploration of natural and wild forests. The flights were carried out in the dry season, weather permitting, at low altitude, around 4 pm when the sun, already low, projected onto the ground shade from the trees whose magnitude informs us perfectly, with a little practice, of the height and therefore the stand age" (Mangin 1933, 650).

22 Arrêté du 24 Octobre 1930 réglementant le régime forestier en Annam.

23 NAV1, RST Service des Forêts 3413: Déclassement partiel de la Réserve No. 292 dite de Dai-Luc, 1937.

24 The dominance of thinking on the links between hydrology and forest cover can be seen elsewhere in Asia; "many Dutch foresters in the Dutch East Indies were greatly influenced by the belief that the presence of forest was the only significant influence

over water distribution (hydrological condition). . . . However, little quantitative data was provided to support this argument and it seems likely that policy preference was in fact the driving factor" (Galudra and Sirait 2009, 525). But it was the French who were perhaps most wedded to these ideas, given the strong thinking in the metropole associating Alpine flooding with lack of forest cover.

25 As Forsyth and Walker point out in their analysis of the use of the same metaphor in Thailand, "The beauty of the 'forest as sponge' narrative, and a key factor in its persistence, is that it can explain both excess water (in the form of flooding) and insufficient water (in the form of dry-season water shortage). There is rarely a year when the occurrence of one (or both) of these environmental events does not 'confirm' the 'truth' of the negative impact of upland forest clearing" (Forsyth and Walker 2008, 114). Not all scientists in Indochina accepted the dominance of the role of forest cover in regulating water supplies, however. For example, Pierre Gourou argued, "Not that the forest exerts the favorable influence on the climate and the regime of the rivers that some have believed should be attributed to it. The forest does not cause noticeably more abundant and more regular rains; it does not effectually lessen the run-off of rain water, and does not diminish the river's floods. In fact, in a country of violent and prolonged rains like French Indochina the retaining canopy of leaves, trunks and roots is rapidly exceeded" (Gourou 1940, 491–92).

26 For example, Henri Lecomte, author of *Les Bois Coloniaux*, wrote in 1923, "I want to specifically make a wish. In the most urgent manner, I call for the creation, in our vast tropical forest area, of 'botanical reserves' which would substantially differ from what is commonly called the 'Forest Reserves.' These are simply areas reserved methodically for later operations. In the 'botanical reserves' that I advocate, these areas would remain indefinitely free from any exploitation. It is clear that forest operations, regulated or not, remove from the primeval forest a multitude of trees and shrubs that do not reappear in the secondary forest. But our knowledge of the rainforest is still very incomplete…Destroying the primeval forest will perhaps expose to extinction interesting plants, which have so far escaped the knowledge of travelers and naturalists. In order to arrange opportunities for the conservation of these plants, perhaps unknown at present, I solicit the creation of 'botanical reserves.'…for the benefit of the generations that follow ours" (Lecomte 1923, in Thomas 2009, 119).

27 NAV1 RST 75.430: La lutte contre les feux de brousse dans les régions de Dalat et la province du Haut-Donnai, 1935.

28 NAV1, RST 56.834: Mesures de protection des forêts contre la pratique du ray et contre les feux de brousse au Tonkin.

29 Ibid.

30 NAV1, RST 78.384: Letter from Hòa Bình Résident to RST, 10 July 1907, A.s. des rays.

31 The fact that they did not do so demonstrates to historian Frédéric Thomas that "the emotion aroused by the destruction of primary forest . . . is not the loss of floristic richness or rupture of an ecosystem but the disappearance of species with high added value destroyed by the clearing. Under these conditions, the environmental argument against ray is a spurious justification that aims to hide the commercial purpose of the fight against this practice" (Thomas 1998, 81).

32 For example, the Résident of Lao Kay fought against local Forest Service regulations

that he was asked to implement in 1916, successfully arguing that swidden could not and should not be banned, but that instead local people could be instructed "in proper burning locations and techniques to preserve regeneration." CAOM, RST NF 4075: Régime foncier et practique des rays à Sơn La; Rapport sur la mise en œuvre agricole de la Province de Sơn La.

33 NAV1, RST N93 75.450: Pratiqué des 'rays,' 1931–39.

34 NAV4, RSA/HC 1086: Rapport de gestion Année 1919 Service Forestier, p. 24.

35 Ibid, p. 24.

36 NAV1, RST 75.386: Création des postes forestiers dans diverses provinces 1925–33. *Quốc Ngữ* refers to the Latin alphabet for the Vietnamese language, and posters with forest rules and regulations were often provided in both the Latin and character-based alphabets in use at the time.

37 NAV1, RST 75.386: Création des postes forestiers dans diverses provinces, 1925–33.

38 NAV4, RSA/HC 3216: Requête des habitants des villages de Trung Dao and Truoc Ha (Quảng Nam) a.s. des exploitations de resins, 1934.

39 CAOM, GGI 65518: Commission d'enquête sur les évènements du Nord-Annam, folder 7F38 (6): Rapport du Chef de Bataillon Garnier sur les causes du mouvement insurrectionnel dans la province de Ha Tinh et sur les remèdes qui pourraient y être apportés, 12 July 1931, p. 35.

40 Ibid., p. 36.

41 CAOM, GGI 65518: Commission d'enquête sur les évènements du Nord-Annam, 7F38: Note de son excellence Tôn Thất Trâm, membre de la commission étude Annam.

Chapter 2. Planting New People

1 In Vietnamese, "*Vì lợi ích mười năm thì phải trồng cây, vì lợi ích trăm năm thì phải trồng người*".

2 Policies for migration, sedentarization, land planning, and agricultural development projects have long been used to incorporate marginal areas into the state, whether the state was colonial or postcolonial, socialist or capitalist. These processes have been termed the envelopment of "non-state spaces" to turn them into "state spaces" (Scott 2009), or as "civilizing projects" (Harrell 1995).

3 Luật Cải cách Ruộng đất [Law on land reform], Dec. 1953.

4 Nghị định 05-NL-QT-NĐ 1956: Quy định việc phân loại rừng, việc khai thác gỗ, củi

5 NAV3, Uỷ ban Khoa học Nhà nước files, folder 4778: Báo cáo sơ bộ về công tác điều tra nghiên cứu thiên nhiên miền bắc Việt Nam (đến hết năm 1959).

6 NAV3, Bộ Nông lâm files, folder 5805: Kế hoạch, báo cáo sản xuất gỗ xuất khẩu trong 1956.

7 NAV3, Bộ Nông lâm files, folder 5582: Rừng Việt Nam với nhiệm vụ kiến thiết trong hoà bình, not dated, likely 1955.

8 Ibid.

9 Thông tư 1303 BCN/VP năm 1946 về sự ích lợi của rừng và sự cần thiết bảo vệ lâm phận quốc gia do Bộ Nội vụ và Bộ Canh nông ban hành.

10 Thông tư 04-NL-TT năm 1956 về việc áp dụng thể lệ khai thác gỗ củi do Bộ Nông nghiệp ban hành.

11 Chỉ thị 335-TTg năm 1959 về công tác lâm nghiệp do Thủ tướng Chính phủ ban hành.
12 NAV3, Bộ Nông nghiệp và Phát triển Nông thôn files, folder Nguyên tác Tổng cục Lâm nghiệp: Cần đẩy mạnh công nghiệp khai thác gỗ năm 1962.
13 *Lâm* and *rừng*, which both mean "forest," are not used interchangeably in Vietnamese. *Lâm* is a Sino-Vietnamese loan word, from the Chinese character 林 (lín). *Lâm* is nearly always used as a modifier (unless it is used as a proper name), such as in "*lâm nghiệp*" (literally "forest work," or forestry) or *lâm sản* (forest product). *Rừng* is a native Vietnamese word, likely a distant loan from the ethnic minority Mường language (Phan 2010). *Rừng* is always a noun; for example, *rừng tự nhiên* (natural forest) or *rừng rậm* (jungle). *Rừng* often implies more wilderness than do compounds of *lâm*, e.g., *đất lâm nghiệp* (forestry land) or *lâm viên* (forest park), both of which imply human management.
14 Thông tư 37-NL/LN năm 1959 về việc lập quy hoạch các vùng kinh tế lâm nghiệp do Bộ Nông lâm ban hành.
15 NAV 3, Tổng cục Lâm nghiệp files, folder 3 TCLN/KT July 23, 1960: Chỉ thị về biện pháp thực hiện kế hoạch quý III năm 1960.
16 NAV3, Tổng cục Lâm nghiệp files, folder 56: Chỉ thị về việc bảo vệ rừng, October 3, 1970.
17 NAV3, Tổng cục Lâm nghiệp files, folder LT 26: Kế hoạch hướng dẫn thực hiện chế độ trả lương theo sản phẩm, June 26, 1961.
18 NAV3, Bộ Nông nghiệp và Phát triển Nông thôn files, folder 30 KH: Về điều tra quy hoạch đường Cúc Phương, June 1963.
19 Eventually, employment at SFEs became more specialized, and recruits with training in forestry were sought out. A unified forestry university was first founded in 1964, when departments from the Agro-Forestry School, now Hanoi Agricultural University, were separated. The university was located in Quảng Ninh province in the country's northeast for many years, before eventually relocating outside of Hanoi in Xuân Mai in 1984. In addition, Cúc Phương National Park south of Hanoi was originally set up in 1962 as a training site for students in forestry practices.
20 NAV 3, Bộ Nông lâm files, folder Lâm trường 26: Kế hoạch hướng dẫn thực hiện chế độ trả lương.
21 One example: "wasteful" and "superstitious" practices such as elaborate wedding and funeral feasts were to be abolished and replaced with more civilized and equitable customs (Lương 1992; Malarney 2002).
22 NAV 3, Bộ Nông Lâm files, April 4 1960: Chỉ thị phát động đợt thi đua chào mừng nước Việt Nam dân chủ cộng hòa 15 tuổi.
23 Interview, Buôn Mê Thuột, Đắk Lắk province, February 2001.
24 In a Swedish evaluation to determine whether SFEs had used "compulsion" to motivate labor migration (and which therefore might be in violation of Swedish law, given that the country was providing forestry aid at this time), researchers found that volunteerism was the primary motivating factor for SFE workers to move, and that given there was no shortage of labor, compulsion was not necessary (Liljeström et al. 1988).
25 As these women traveled to work at SFEs, they did not always give up their dreams of forming families of their own. Some were able to find male partners among fellow SFE workers. But for other women, they resorted to single motherhood in order to fulfill

their dreams of having children (Phinney 2005). The phenomenon of single mothers at SFEs was so serious that several workshops on the issue, led by gender specialists and sociologists, were held in the 1980s (Trung tâm Nghiên cứu Phụ nữ 1987).

26 The highest position a woman ever attained was as deputy director of a subdepartment at an SFE (Nghiêm Thị Yến 1994).

27 The advice from the cadres was to choose fields that were not too steep and had good soil; to concentrate fields together in order to protect crops from forest animals; to make swidden in *Imperata* fields where it would not compete with timber production; and to improve technology in swidden fields, such as using natural and chemical fertilizers. NAV3, Bộ Nông lâm files, folder 5628: Báo cáo công tác lâm nghiệp trong 1955 của Ty Nông lâm.

28 NAV3, Bộ Nông nghiệp và Phát triển Nông thôn files, folder TCLN 14 LN/VP: Chỉ thị về phương hướng công tác lâm nghiêp khu tư trị Tây Bắc, March 20, 1963.

29 For example, one author argued in the pages of *Vietnamese Studies* that shifting cultivation fostered primitive cultural habits and harmful customs; created unhealthy and disease-ridden areas, where people "often drink from thin muddy streams infected with dead leaves and carrions," and added that the practice also "devastates forests, favors floods, and accelerates soil erosion" (An Thu 1968, 202).

30 Resolution of the Government Council No. 38-CP.

31 As anthropologist Nguyễn Văn Chính has noted, there have never been clear definitions of what "nomadic" or "unsettled" peoples did or who they were, but these terms have been widely used since the beginning of the DRV (Nguyễn Văn Chính 2008). Oscar Salemink further points out, "lack of settlement" often referred more to a lack of state legibility: "every village where the state is not present, in whatever form, is considered as not fixed" (Salemink 1997, 506).

32 Interviews, Đakkrông district, Quảng Trị province, September 2005.

33 Interview, Lá To village, Đakkrông district, Quảng Trị province, September 2005.

34 In pointing out defoliation had less of an impact on forests than poor management of SFEs, this is not to say the wartime defoliation was not significant. Indeed, it had profound local effects on the livelihoods of people whose crops and forests were sprayed, and the health consequences of those who live near areas doused with the dioxin-laden chemicals remain serious today (Martini 2012).

35 NAV 3, Tổng Cục Lâm Nghiệp files, folder 120 37 DT/TT: Thông tư hướng dẫn việc tổ chức thực hiện chỉ thị của thủ tướng chính phủ về công tác phân vùng kinh tế nông lâm nghiêp ở miền Nam, November 15, 1975.

36 NAV3 Uỷ ban Khoa học Nhà nước files, folder 4867: Phương hướng điều tra và bảo vệ tài nguyên thiên nhiên trong kế hoạch 1976–80.

37 Interview, SFE Director, Thừa Thiên Huế Province, November 2011.

38 Interview, former head of Forest Inventory and Planning Institute for the South, September 2014.

39 NAV 3, Tổng cục Lâm nghiệp files, folder 120 37 DT/TT: Thông tư hướng dẫn việc tổ chức thực hiện chỉ thị của thủ thướng chính phủ về công tác phân vùng kinh tế nông lâm nghiệp ở miền Nam, November 15, 1975.

40 NAV 3, Tổng cục Lâm nghiệp files, folder 178: 14/CNR: Chỉ thị về đẩy mạnh khai thác vận chuyển cung cấp đảm bảo 2,000,000 m^3 năm 1978, April 13, 1978.

41 NAV 3, Tổng cục Lâm nghiệp files, folder 178: 15/CNR. Chỉ Thị về đẩy mạnh sản xuất gỗ xuất khẩu, April 26, 1978.
42 NAV 3, Tổng cục Lâm nghiệp files, folder 48/VGNN-TLSV: Uỷ bản Vật giá Nhà nước, April 19, 1975.
43 NAV 3, Tổng cục Lâm nghiệp files, folder 4608: November 7, 1975.
44 NAV3, Bộ Nông nghiệp và Phát triển Nông thôn files, folder # 661-TC/LN: Về việc thu tiền lâm sản tại các lâm trường và các Ty Lâm nghiệp QD tự khai thác.
45 NAV3, Bộ Nông nghiệp và Phát triển Nông thôn files, folder 706 LN/TR March 22, 1963: Về phòng chống cháy cây rừng trồng.
46 Interviews, SFE workers in Cẩm Xuyên SFE, June 2001.
47 As one example of the lack of attention to SFEs, the WorldCat library catalogue lists nearly 174 works in Vietnamese on "cooperatives" (*hợp tác xã*) but only three published works specifically on SFEs. There are no English-language works on the SFE era.

Chapter 3. Illegal Loggers and Heroic Rangers

1 NAV3, Bộ Nông nghiệp và Phát triển Nông thôn, folder Bộ Nông nghiệp 56; #55/LN Chỉ thị về việc bảo vệ rừng, [1970].
2 NAV3, Bộ Nông nghiệp và Phát triển Nông thôn, folder Bộ Nông nghiệp 56: # 19 TT-LB Thông tư về việc đẩy mạnh công tác thông tin cổ động phục vụ việc thực hiện các chính sách lớn của Đảng và nhà nước về rừng và bảo vệ rừng.
3 A forest protection unit had been first established in the Department of Agriculture in 1946, with the objective of "achieving justice regarding any cases of forest crimes," but it was in the 1972 Ordinance that the mandate was broadened to include the general work of "protecting the forest" and a modern forest ranger service was born (Cục Kiểm lâm 1998).
4 Interview, Flora and Fauna International, 2005.
5 Many scholars have pointed out that state-classified "forest land" is often land with the least amount of vegetation cover in many communities, and the problem of official criteria for forests not acknowledging local definitions is widespread, particularly in Southeast Asia (Laungaramsri 2000; Wong and Delang 2007).
6 Within a decade from 1990–2000, coffee plantations increased from 50,000 to 500,000 hectares, most of which came from former designated forestry land (Lindskog et al. 2005). Đắk Lắk provincial officials estimated that they lost 40,000 hectares of forest each year in the 1990s; this would account for a loss of half the province's forest cover at the height of the coffee boom (Ngô Đình Quê et al. 2006).
7 Earlier in the 1980s, most forestry aid focused on reforestation and plantations, and primarily came from the UNDP/FAO, the World Food Program (WFP), and Sweden, who had helped Vietnam build a large-scale paper and pulp supply factory, known as Bãi Bằng, in the late 1970s (MOF 1991b).
8 Laws carry more weight than decisions of ministries, as they are passed by the National Assembly.
9 The Comprehensive University had been founded in 1956 after independence to replace the *Université Indochinoise*. In 1993, the Comprehensive University was folded into Vietnam National University-Hanoi. Scientists associated with this college were

some of the only independent voices on environmental issues that were heard in the 1960s and 70s.

10 Interview with Dr. Võ Quý, August 2009.

11 Quyết định 41-TTg năm 1977 Quy định các khu rừng cấm do Thủ tướng Chính phủ ban hành.

12 There were differences between these areas: national parks were for "protecting the value of comprehensive nature protection, scientific research, conservation and cultural monuments, and sightseeing services." Nature reserves were aimed at "forest conservation for the protection of valuable scientific and genetic preservation of animals and plants; reserves can be opened for scientific research but are not open to tourism or cultural needs". Finally, CHERs were designated for "historic, cultural and scenic or aesthetic values of environmental protection, effective for sightseeing, recreation, and leisure." See "Quy chế Quản lý rừng đặc dụng" (Ban hành theo Quyết định số 1171-QĐ ngày 30-12-1986).

13 For documentation of these protected areas' histories, see MOF (1991b) and MARD (1997).

14 From Decree 140/1999/QD-BNN- ĐCĐC.

15 Other studies that have assessed patterns in shifting cultivation with satellite and survey data have shown that there was very little deforestation attributable to shifting cultivation during the *đổi mới* era, as this activity was mostly carried out on previously cleared and now fallowing lands, rather than in primary forests. The expansion of shifting cultivation into new forests almost always took place only when minority villages were pushed to forest margins by Kinh agricultural pioneers (Déry 2003; Meyfroidt et al. 2013).

16 In an interview with a politician who had known Kiệt well, he claimed that the former prime minister wanted a blanket logging ban across the nation, but settled on the regional bans first as a compromise, due to push back from local party elites. Interview with Dr. Nguyễn Đăng Vang, member of the National Assembly and Vice-chairman of Science, Technology and Environment committee, August, 2008.

17 Nghị định 39-CP năm 1994 về hệ thống tổ chức và nhiệm vụ, quyền hạn của Kiểm lâm.

18 Interview, Nguyễn Bá Thụ, Hanoi, August 2008.

19 Interview, Phạm Văn Ngoan, August 1997.

20 Indeed, earlier reincarnations of the *lâm tặc* were nearly celebrated, as was the case in Nguyễn Huy Thiệp's short story, "The Woodcutters" (*Những Người Thợ Xẻ*). In this story, the anti-hero Bường travels the northern mountains offering to cut wood for SFEs (both legally and illegally) and celebrates his life as a "free person" (not employed by the state), although his cheating and sexual licentiousness earn him condemnation in some quarters (Nguyễn Huy Thiệp 2003).

21 Hoàng Cầm relates a similar case in Sơn La, Vietnam, where he conversed with a logger in a Thái ethnic minority village, who noted, "Now the officials and mass media call us forest thieves [*lâm tặc*] and we label ourselves *lâm tặc*, too....But I think," he continued, "we are just little forest thieves [*lâm tặc con*]; the father of forest thieves [*đại ca của lâm tặc*], of course, are the people from the Forest Protection Unit and Kinh traders who are now living in the valley and those who come here to do their timber business" (Hoàng Cầm 2007, 21).

22 In one case in 2003, the local head of the anti-smuggling division of the Ministry of Industry and Trade in Hà Tĩnh was caught smuggling ironwood, while in 2006, the head of KL was accused of providing a letter of permission to an illegal logger who had removed 28 m^3 of valuable timber from a protected watershed forest (DPA 2003; Phan Hồng 2006).

23 Most loggers living in villages around the KGNR sold the timber locally, as no one had access to long distance transport or "big man" connections in my interview sample, and even the best quality timber only fetched around US$200 per m^3 locally within the district market.

24 As other studies have demonstrated, this is a common outcome of national-level criminalization of logging (Humphreys 2008; McCarthy 2002).

Chapter 4. Rule by Reforestation

1 NAV3, Bộ Nông nghiệp và Phát triển Nông thôn files, Văn Phòng Bộ, Folder #1: Cấp thông tư chỉ thị Nghị định của Tổng cục Lâm nghiệp năm 1960.

2 Ibid.

3 NAV 3, Bộ Nông nghiệp và Phát triển Nông thôn files, Liên Bộ Thuỷ Lợi Nông Lâm 2TT/LB: Về việc trồng cây bảo vệ đê.

4 NAV3 Bộ Nông nghiệp và Phát triển Nông thôn files: 6 TD/LN Chỉ thị về phát động Tết Trồng Cây quyết thắng giặc Mỹ xâm lược.

5 Ibid.

6 Interview, Maurice Williamson, Oct 2000.

7 The official title of the program was the "Master Guidelines and Policies to Utilize Unoccupied Land, Barren Hilly Areas, Denuded Land and Beaches and Waterfront."

8 Households who protected existing forest received on average 50,000 VND/ha (approximately US$4/ha) payments, and those who reforested got an initial investment of 1.5 million VND per hectare (US$125) with yearly payments of up to 300,000 VND (US$25) for maintaining this land. There were also funds for sedentarization, around 800,000 VND (US$66) per household, for those who agreed to move out of protected watershed forests (Nguyễn Vành et al. 1995).

9 A later review of the 327 program confirmed, "Projects are formulated at commune or district level and follow centrally established guidelines. There is seldom consultation with local people—the program's supposed target groups—by Government staff who many not appreciate the broad constraints impacting on poor rural households. Having been developed in isolation, project objectives and targets often do not reflect poor rural households' reality. Local people, the keys to project success, have no ownership of project objectives" (World Bank 1998, 11).

10 This idea that any tree is good has been labeled "forest fundamentalism" by Michael Dove (2004, 426) and an example of an "Edenic narrative" (Cohen 1999).

11 Yet not all donors entirely believed that it was feasible or realistic to try to raise forest cover, or that the plans would be particularly environmentally sound. A consultant to the Swedish aid agency, SIDA, told me "I don't see why the government thinks we should have 43 percent forest cover like in 1943. We should have enough to meet water needs and subsistence and no more." An Australian forestry advisor was appalled that

AusAid was contributing funding to a project he considered ill-advised and a waste of money, questioning, "Why are they planting so many bloody trees, and in such bloody bad places?"

12 Historian Frédéric Thomas has argued that assertions about the exact extent of forest cover in 1943 were unlikely to be entirely accurate given the poor inventories at the time, and reflect more of the political nature of ideas of forest cover and the difficulties of establishing clear concepts like deforestation (Thomas 2000, 49).

13 As a recent review of forest plantations in Asia has noted, "the act of defining an area of forestland as 'degraded' is fundamentally a political act which shapes decisions regarding land-use and access" (Barr & Sayer 2012, 10). Such classifications of bare lands and grasslands as deficient are particularly common in Southeast Asia, as Michael Dove has noted (Dove 2004).

14 As Thomas Sikor has pointed out, "land is classified as barren if its actual land use does not coincide with its designated use, hence does not meet its function in national development planning" (Sikor 1995, 147).

15 Similar top-down approaches were reported in other areas, where farmers were "merely given the red book by their hamlet chief and told it was their forest land. Yet they have no knowledge of how the plots were selected or how the book can be used. The red books, in fact, have different meanings for the local people than they do for the state" (Sowerwine 2004, 109).

16 A similar situation is related in Sikor and Nguyễn Quang Tân (2007).

17 As many authors have noted, tree planting is one of the most common ways to secure land tenure, as the tree serves as a visual anchor object establishing active property rights (Cernea 1992).

18 As the authors of a study of land rights in Southeast Asia note, exclusion is a double-edged sword, as it is both a necessary condition for land use as well as a source of conflict; in this, "exclusion creates both security and insecurity" (Hall et al. 2011, 8).

19 This is not to say that protests did not occur around the KGNR as exclusions happened. Some of the earliest claimants of bare hill land who had reforested with 5MHRP seedlings lamented that up to 30 percent of their trees had been stolen rather brazenly ("*vô tư*") by other neighbors. One man who had claimed bare hill land complained that people who had previously harvested fuelwood there continued to do so—this time taking his newly planted trees in secret, which he wanted to denounce and expose ("*vạch mặt*"), but he could not get the neighbors who lived around the land he claimed to tell him who was taking the trees.

20 Authors working in upland minority areas of Hòa Bình province noted that farmers' swiddening practices had long been pointed to by officials as a disastrous cause of forest loss and subsequent hydrological problems: "Farmers were accused of being responsible for an alleged ecological disaster. They were singled out as the guilty ones and, following this argumentation, it was logical in people's consciousness that they had to atone for their faults by reforesting the hills. Farmers strongly believe that runoff from the watershed increases with forest cover. This belief is so entrenched in people's minds that some farmers use it to explain all land management problems" (Clement and Amezaga 2008, 272)

21 For example, Rick Schroeder has analyzed the situation in The Gambia, where a donor

aid project gave seedlings out to promote agroforestry, and men often planted these trees in women's vegetable gardens. What might have been seen by technical planners as an eager adoption of the tenets of intercropping and agroforestry was in fact a way for the men to gain control of women's rights to land and income (Schroeder 1999).

22 Territorialization is defined as the control, mapping, or administration of land, labor and resources by the state or other social power (Sack 1986, 1), including appropriating and assigning names and classifications to people and places (Worby 1994; Vandergeest and Peluso 1995); moving and resettling populations to land frontiers (De Koninck and Déry 1997); or mapping borders (Winichakul 1994).

Chapter 5. Calculating Carbon and Ecosystem Services

1 The acronym REDD+ is to indicate the additionality of "the role of conservation, sustainable management of forests, and enhancement of forest carbon stocks in developing countries."

2 Interview, Ministry of Environment and Natural Resources, March 10, 2009.

3 This is also known as Decision No. 380 QĐ-TTG.

4 Vietnam's list of compensated services is closest to the services funded by Costa Rica's national PES policy of: 1) greenhouse gas mitigation; (2) hydrological services; (3) scenic value; and (4) biodiversity, indicating that the visits from Costa Rican officials may have particularly influenced this policy (Sanchez-Azofeifa et al. 2007).

5 Due to slowness in developing the rules by which targeted buyers must comply with PES, hydropower companies are currently the most regulated entities (235 companies are contracted to pay PES fees as of 2014), with a smaller percentage of domestic water supply companies (72 in total, while many more do not yet participate because watersheds that supply their water have not yet been mapped). No industrial water users currently participate, and PES contracts have been negotiated with around 44 tourism companies (located in only four participating provinces) out of thousands in Vietnam. See McElwee and Nguyễn Chí Thanh (2015).

6 Interview, Phạm Hồng Lượng, director of Vietnam Forest Development and Protection Fund, August 12, 2014.

7 The practice of socialization has meant radical change from the days of cooperatives and collective state enterprises that provided housing, medical, and educational care for workers. After just a decade of *đổi mới*, for example, the average citizen paid out-of-pocket for most social services; these private expenditures accounted for about 70 percent of the country's education costs and 80 percent of health costs (London 2004).

8 Nguyễn Bá Ngãi, vice-director of VNFOREST, in discussion with REDD Network, Hanoi, November 2011.

9 Interview, head of Hoà Bình Department of Agriculture and Rural Development, August 2014.

10 SWAT was first developed in the US by the Department of Agriculture and Texas A&M University to "predict the impact of land management practices on water, sediment and agricultural chemical yields in large complex watersheds with varying soils, land use and management conditions over long periods of time" (Neitsch et al. 2001, 1), and was built using data from the arid and agriculturally dominant southwest

US. Despite its narrow origins, this model has been adopted in many international situations, and ARBCP consultants used it to predict how the highland pine forests of central Vietnam would produce water for hydroelectric systems, and how much less water might be produced if these lands were converted to agriculture.

11 The power output of the Đa Nhim hydropower plant is 1 billion kwh/year, which requires around 550 million m3 water to flow through turbines downstream from the reservoir. Đa Nhim reservoir has a capacity of 165 million m3 water storage; therefore, it was estimated that nearly 400 million m3 water is held and released through the forests that surround the reservoir in the watershed.

12 The fact that national fee levels have been set based on very narrow local studies was a source of concern for many people interviewed. In fact, at the Đa Nhim hydropower company (the original model for the valuation of hydrological services in the PES pilots), an engineer pointed out that the valuation referred only to downstream water flow, and not to the efficiency of hydropower production. Different hydropower plants use water more or less efficiently, and the engineer stated that his particular plant used much less water per kwh produced, yet received no reduction in its fees compared to a plant that used twice as much water to produce the same energy. Interview, November 2011.

13 Indeed, MARD encouraged provinces to focus only on forest cover as a proxy for environmental services through Circular 60/2012/TT-BNNPTNT, dated 09/11/2012, "Regulation on principles to identify forest areas in the watersheds providing payments for forest environmental services."

14 Interview, Thừa Thiên Huế province Forest Protection and Development Fund, August 2014.

15 Interview, Cát Tiền National Park, August 2014.

16 Interview, Bù Gia Mập National Park, August 2014.

17 Interview, Lào Cai Province, August 2014.

18 Win-win scenarios have been promoted by donors about households receiving hundreds of dollars a year from PES, but these were based on the assumption that households would have large plots of land that for which they will receive payment, given the relatively low amount (less than $20/ha) paid from PES in most areas so far. But as noted in the previous chapter, households only control 31 percent of the total forest estate, and of that percentage, census results indicate that only 8.5 percent of households who have red books for forestry have plots larger than five hectares. The nation-wide average of forest land per household is only 2.91 ha, and many millions of households have no holdings at all (McElwee 2012).

19 These details were laid out in Circular 80/2011/TT-BNN, dated 23/11/2011, from MARD.

20 Yet it was not just households who saw PES as more of the same type of state subsidy policy for forestry. During interviews, officials from several state forest enterprises in Lâm Đồng emphasized that, to them, the PES policy was simply the same as the 5MHRP reforestation policy that had come before it; the latter had only been repackaged with a new name. Both projects amounted to money coming from outside that went to provincial entities and allowed them to continue to exist and pay their workers' salaries, even if one (PES) was supposed to be a market policy, not a state subsidy policy. To the SFEs, these policies were one and the same.

21 Interview, vice-director of Bi Đúp Núi Bà National Park, November 2011.

22 Also known as Decision No.799/QĐ-TTg of the Prime Minister on June 27, 2012.

23 For example, Kẻ Gỗ Nature Reserve is one of many sites that have been targeted by multiple REDD+ projects, including a project titled "Advancing Understanding of Forest Carbon Stock Enhancement" funded by the Dutch NGO SNV, as well as a project on "Sustainable Management of Forests through Low-Emissions Development Planning for Green Growth in the North Central Agro-Ecological Region" to be funded by the World Bank's Forest Carbon Partnership Facility.

24 These calculations produce new entities—such as "a sum, an ordered list or an evaluation that corresponds to what has been carried out in the calculative space and that links and summarizes the entities taken into account. The resulting entity is not new in the sense of springing from nowhere, but it 'has to be able to leave the calculative space and circulate elsewhere in an acceptable way (without taking along all the calculative apparatus)'" (Callon and Muniesa (2005, 1231) in Lövbrand & Stripple (2011, 189).

25 Known as Decision 2855/BNN-KHCN, September 17, 2008.

26 As of 2014, estimates of what forest carbon will cost on a REDD+ market range between $5 and $10 a ton, based on existing voluntary forest carbon markets (World Bank 2014).

27 I am particularly grateful to Bernhard Huber for raising this important point.

28 As one study has noted, "The wide range of reported [carbon] stocks (25–143 Mg/ha) associated with mature rubber stands in Southeast Asia demonstrate that this tree-based land cover in some environments could sequester substantially less carbon than various types of mature swidden fallows" (Fox et al. 2014, 322). (Mg equals one tonne of carbon.)

Conclusion

1 News reports on the 6,700 tree protest can be found in Đăng Minh (2015) and Thúy Hạnh (2015).

2 As one example, recent work by forest ecologists has even asserted that individual trees may "speak" to one another regarding nutrient access by way of fungal networks embedded in tree roots (Messier et al. 2013).

References

Archival Collections

Centre des Archives d'Outre-mer, Aix-en-Provence, France (CAOM)
Fonds de la Résidence Supérieure du Tonkin, Nouveau fonds (RST NF)
Fonds de la Résidence Supérieure en Annam (RSA)
Fonds du Gouvernement Général de l'Indochine (GGI)
Indochine Nouveau Fonds (INF)
National Archives of Vietnam (Trung Tâm Lưu Trữ Quốc Gia) Number 1, Hanoi
Fonds de la Résidence Supérieure au Tonkin (RST)
Fonds de la Service des Forêts
National Archives of Vietnam (Trung Tâm Lưu Trữ Quốc Gia) Number 2, Ho Chi Minh City
Fonds du Gouvernement de la Cochinchine
National Archives of Vietnam (Trung Tâm Lưu Trữ Quốc Gia) Number 3, Hanoi
Bộ Nông lâm (Ministry of Agriculture and Forestry) folders
Bộ Nông nghiệp và Phát triển Nông thôn (Ministry of Agriculture and Rural Development) folders
Tổng cục Lâm nghiệp (General Directorate of Forestry) folders
Uỷ ban Khoa học Nhà nước (State Committee on Science) folders
National Archives of Vietnam (Trung Tâm Lưu Trữ Quốc Gia) Number 4, Dalat
Fonds de la Résidence Supérieure de l'Annam (RSA/HC)

French Colonial Legal Documents

Arrêté réglementant le service forestier en Cochinchine, June 23, 1894
Arrêté du Gouverneur Général du 3 juin 1902 Concernant l'exploitation des produits forestiers au Tonkin
Arrêté du 5 Sept 1905 réglementant les coupes en domaine forestier réservé
Arrêté du 19 Mars 1907 concernant la constitutions de réservé forestier
Arrêté du Gouverneur Général concernant les bois et forêts sis sur le territoire de la Cochinchine, 1913
Régime Forestier en Annam, 1926
Arrêté du 24 Octobre 1930 réglementant le régime forestier en Annam

Laws Passed by the Vietnamese National Assembly

Luật Cải cách Ruộng đất [Law on Land Reform], Dec. 1953
Luật Bảo vệ và Phát triển Rừng [Law on Forest Resources Protection and Development], 1991
Luật Đất Đai [Land Law], 1993

Vietnamese Prime Ministerial and Government Council Policies

Chỉ thị 335-TTg năm 1959 về công tác lâm nghiệp do Thủ tướng Chính phủ ban hành [Directive 335-TTg on forestry work issued by the prime minister]
Nghị định 05-NL-QT-NĐ 1956 Quy định việc phân loại rừng, việc khai thác gỗ, củi [Decree 5 regulations on forest classification, exploitation of timber and firewood]
Nghị định 39-CP năm 1994 về hệ thống tổ chức và nhiệm vụ, quyền hạn của Kiểm lâm [Decree 39 on the system of organization of duties and rights of the Forest Ranger service]
Nghị định 99/2010/NĐ-CP về chính sách chi trả dịch vụ môi trường rừng chính phủ [Decree 99 on the policy for payment for forest environmental services]
Nghị quyết của Hội đồng Chính phủ 38/CP on 12/3/1968 về cuộc vận động định canh định cư [Resolution 38/CP of the Government Council regarding the mobilization of sedentarization]
Quyết định 140/1999/ QĐ-BNN-ĐCĐC về nội dung, tiêu chí định canh định cư [Decision 140 on the content and criteria for sedentarization]
Quyết định 1171-QĐ ngày 30-12-1986 quy chế quản lý rừng đặc dụng [Decision 1171 on the regulations on special-use forests]
Quyết định 41-TTg năm 1977 quy định các khu rừng cấm do Thủ tướng Chính phủ ban hành [Decision 41 on regulating the prohibited forests, issued by the prime minister].
Quyết định No.799/QĐ-TTg do Thủ tướng Chính phủ ban hành Phê duyệt Chương trình hành động quốc gia về "Giảm phát thải khí nhà kính thông qua nỗ lực hạn chế mất rừng và suy thoái rừng, quản lý bền vững tài nguyên rừng, bảo tồn và nâng cao trữ lượng các bon rừng" giai đoạn 2011–2020 [Decision 799 on the prime minister's approval of Vietnam's REDD+ National Action Plan for 2011–20]
Quyết định 2855/BNN-KHCN 17/9/2008 công bố việc xác định cây cao su là cây đa mục đích [Decision 2855 declaration on determining that rubber is a multipurpose tree]
Quyết định 327-CT về một số chủ trương, chính sách sử dụng đất trống, đồi núi trọc, rừng, bãi bồi ven biển và mặt nước chủ tịch hội đồng bộ trưởng [Decision 327–CT of the Government Council on master guidelines and policies to utilize unoccupied land, barren hilly areas, denuded land and beaches and waterfronts]
Quyết định 661/QĐ-TTg về mục tiêu, nhiệm vụ, chính sách và tổ chức thực hiện Dự án trồng mới 5 triệu ha rừng [Decision 661on the objectives, responsibilities, policies and organization for the implementation of the 5 Million Hectare Reforestation Program]
Quyết định số 380/QĐ-TTg 10/4/2008 của Thủ tướng Chính phủ về chính sách thí điểm chi trả dịch vụ môi trường rừng [Decision 380 QĐ-TTG of the prime minister on the pilot policy for payments for forest environmental services]

Policies Issued by Vietnamese Ministerial-Level Departments

Thông tư 1303 BCN/VP năm 1946 về sự ích lợi của rừng và sự cần thiết bảo vệ lâm phận quốc gia do Bộ Nội vụ và Bộ Canh nông ban hành [Circular 1303 on the benefits of forests and the necessity of protecting national forests of the Ministry of Agriculture]

Thông tư 37-NL/LN năm 1959 về việc lập quy hoạch các vùng kinh tế lâm nghiệp do Bộ Nông lâm ban hành [Circular 37 on the work of planning forest economic areas of the Minister of Agriculture and Forestry]

Thông tư 04-NL-TT năm 1956 về việc áp dụng thể lệ khai thác gỗ củi do Bộ Nông nghiệp ban hành [Circular 4 on the adoption of a code for logging and fuelwood collection of the Ministry of Agriculture]

Thông tư 60/2012/TT-BNNPTNT nguyên tắc phương pháp xác định diện tích rừng [Circular 60 on regulation on principles to identify forest areas in the watersheds providing payments for forest environmental services of the Ministry of Agriculture and Rural Development]

Thông tư 80/2011/TT- BNNPTNT hướng dẫn phương pháp xác định tiền chi trả dịch vụ [Circular 80 on methods guiding identification of payments for forest environmental services of the Ministry of Agriculture and Rural Development]

Thông tư số 34/2009/TT-BNNPTNT quy định tiêu chí xác định và phân loại rừng [Circular 34 on forest classification of the Ministry of Agriculture and Rural Development]

Books and Articles

Agarwal, Anil. 1984. "Vietnam after the Storm." *New Scientist*, 10 May: 10–14.

Agrawal, Arun. 2005. *Environmentality: Technologies of Government and the Making of Subjects*. Durham: Duke University Press.

Agrawal, Arun, and Maria Lemos. 2007. "A Greener Revolution in the Making? Environmental Governance in the 21st Century." *Environment: Science and Policy for Sustainable Development* 49 (5): 36–45.

Ahlback, Arnold. 1995. "On forestry in Vietnam, the new reforestation strategy and UN assistance." *Commonwealth Forestry Review* 74 (3): 224–229.

Akrich, Madeleine, Michel Callon and Bruno Latour. 2002. "The Key to Success in Innovation Part II: The Art of Choosing Good Spokespersons." *International Journal of Innovation Management* 6 (2): 207–25.

Allonard, P.M. 1937. *Pratiqué de la lutte contre les feux de brousse: surveillance de la population, moyens de propagande, lignes pare-feux* [Practicing the fight against bushfires: Population monitoring, propaganda, firewalls]. Hanoi: Imprimerie d'Extrême-Orient.

An Thu. 1968. "The Zao Are Coming Down to the Lowlands." *Vietnamese Studies* 15: 175–87.

Anon. 1968. "Vị trí của rừng núi trong công cuộc xây dựng chủ nghĩa xã hội ở nước ta" [Position of mountain forests in the work of constructing socialism in our country]. *Học Tập* [Academic Journal] 8: 34–39.

Anon. 2007. "Kẻ Gỗ: tạm qua những ngày 'nóng'" [Ke Go: "Hotspot" Days]. *ThienNhien.Net*. http://www.thiennhien.net/2007/08/14/ke-go-tam-qua-nhung-ngay-nong/.

Anon. 2011. "Lawmakers Doubt Government's Afforestation Figures." *Tuổi Trẻ* [Youth], October 13, p. 1.

ARBCP. 2009. "Alternative Livelihoods for Improved Biodiversity Conservation." Presentation by Asia Regional Biodiversity Conservation Program at South-East Asia Workshop on Payments for Ecosystem Services, Bangkok, Thailand, June 29–July 1.

Armitage, Ian. 1990. *Forestry Sector Review Tropical Forestry Action Plan Vietnam: Management of Natural Forests.* Hanoi: Ministry of Forestry, UNDP, and FAO.

Ascher, William. 1999. *Why Governments Waste Natural Resources: Policy Failures in Developing Countries*. Baltimore: The John Hopkins University Press.

Ashton, Peter. 1986. "Regeneration in Inland Lowland Forests in South Viet-Nam One Decade after Aerial Spraying by Agent Orange." *Bois et Forêts des Tropiques* [Tropical Wood and Forests] 211: 1–30.

Aso, Mitch. 2011. "Profits or People? Rubber Plantations and Everyday Technology in Rural Indochina." *Modern Asian Studies* 46 (1): 19–45.

AwBeng, Teck. 1997. "Vanishing Forests in Vietnam." *Asian Timber* 16 (7): 22–26.

Baillaud, Emile. 1914. "Le régime forestier dans les colonies françaises" [The forest regime in the French colonies]. In *Le régime forestier aux colonies* [Colonial forest regimes], vol. 3, pp. 1–14. Bruxelles: L'Institut Colonial Internationale.

Bakker, Karen. 2005. "Neoliberalizing Nature? Market Environmentalism in Water Supply in England and Wales." *Annals of the Association of American Geographers* 95 (3): 542–65.

Bamford, Paul. 1955. "French Forest Legislation and Administration, 1660-1789." *Agricultural History* 29 (3): 97–107.

———. 1956. *Forests and French Sea Power, 1660–1789*. Toronto: University of Toronto Press.

Barney, Keith. 2008. "Local Vulnerability, Project Risk, and Intractable Debt: The Politics of Smallholder Eucalyptus Promotion in Salavane Province, Southern Laos." In *Smallholder Tree Growing for Rural Development and Environmental Services*, edited by Denyse Snelder and Rodel Lasco, 263–86. Dordrecht: Springer.

Barnaud, Cecile, and Martine Antona. 2014. "Deconstructing Ecosystem Services: Uncertainties and Controversies around a Socially Constructed Concept." *Geoforum* 56: 113–23.

Barr, Christopher, and Jeffrey Sayer. 2012. "The Political Economy of Reforestation and Forest Restoration in Asia-Pacific: Critical Issues for REDD." *Biological Conservation* 154: 9–19.

Bassford, John. 1987. "The Franco-Vietnamese Conception of Land Ownership." In *Borrowings and Adaptations in Vietnamese Culture*, edited by Trương Bửu Lâm, 84–99. Honolulu: Center for Asian and Pacific Studies, University of Hawai'i.

BBC. 2015. "Vietnamese Push Back on Facebook to Save Hanoi's Trees." *BBC Trending* blog, March 24. http://www.bbc.com/news/blogs-trending-31991940.

Beresford, Melanie and Lyn Fraser. 1992. "Political Economy of the Environment in Vietnam." *Journal of Contemporary Asia* 22 (1): 3–19.

Bertin, Andre. 1924. *Notes sur les bois de l'Indochine* [Notes on the Woods of Indochina]. Paris: Agence Économique de l'Indochine.

Bevir, Mark. 2010. "Rethinking Governmentality: Towards Genealogies of Governance." *European Journal of Social Theory* 13 (4): 423–41.
Bhattarai, Madhusudan, and Michael Hammig. 2001. "Institutions and the Environmental Kuznets Curve for Deforestation: A Cross-country Analysis for Latin America, Africa and Asia." *World Development* 29 (6): 995–1010.
Biggs, David. 2005. "Managing a Rebel Landscape: Conservation, Pioneers, and the Revolutionary Past in the U Minh Forest, Vietnam." *Environmental History* 10 (3): 448–76.
———. 2008. "Breaking from the Colonial Mold: Water Engineering and the Failure of Nation-Building in the Plain of Reeds, Vietnam." *Technology and Culture* 49 (3): 599–623.
———. 2012. *Quagmire: Nation-Building and Nature in the Mekong Delta.* Seattle: University of Washington Press.
Binh, T. N. K. D., Nico Vromant, Nguyen Thanh Hung, Luc Hens, and E.K. Boon. 2005. "Land Cover Changes Between 1968 and 2003 in Cai Nuoc, Ca Mau Peninsula, Vietnam." *Environment, Development and Sustainability* 7 (4): 519–36.
Birkenholtz, Trevor. 2009. "Irrigated Landscapes, Produced Scarcity, and Adaptive Social Institutions in Rajasthan, India." *Annals of the Association of American Geographers* 99 (1): 118–37.
Blaikie, Piers. 2008. "Epilogue: Towards a Future for Political Ecology that Works." *Geoforum* 39: 765–72.
Blaikie, Piers and Harold Brookfield. 1987. *Land Degradation and Society.* London: Routledge Kegan & Paul.
Bộ Nông Lâm [Ministry of Agriculture and Forestry]. 1957. *Bảo vệ rừng trong sản xuất nương rẫy* [Protecting forests in swidden production]. Hanoi: Ministry of Agriculture and Forestry.
Bose, Purabi, Bas Arts, and Han van Dijk. 2012. "'Forest Governmentality': A Genealogy of Subject-Making of Forest-Dependent 'Scheduled Tribes' in India." *Land Use Policy* 29 (3): 664–73.
Braun, Bruce. 2004. "Querying Posthumanisms." *Geoforum* 35 (3): 269–73.
———. 2006. "Environmental Issues: Global Natures in the Space of Assemblage." *Progress in Human Geography* 30 (5): 644–54.
Brenier, Henri. 1914. *Essai d'Atlas statistique de l'Indochine française.* Hanoi: Gouvernement général de l'Indochine.
Brocheux, Pierre, and Daniel Hémery. 2011. *Indochina: An Ambiguous Colonization, 1858–1954.* Berkeley: University of California Press.
Brockington, Daniel. 2011. "Ecosystem Services and Fictitious Commodities." *Environmental Conservation* 38 (4): 367–69.
Brouwer, Roland. 1995. *Planting Power: The Afforestation of the Commons and State Formation in Portugal.* Delft: Eburon.
Brown, George. 1957. *The Forests of Free Viet-Nam: A Preliminary Study for Organization, Protection, Policy, and Production.* Saigon: USOM/Saigon.
Bruun, Thilde, Andreas Neergaard, Deborah Lawrence, and Alan Ziegler. 2009. "Environmental Consequences of the Demise in Swidden Cultivation in Southeast Asia: Carbon Storage and Soil Quality." *Human Ecology* 37 (3): 375–88.

Bryant, Raymond. 1994. "Shifting the Cultivator: The Politics of Teak Regeneration in Colonial Burma." *Modern Asian Studies* 28 (2): 225–50.

Buchy, Marlene. 1993. "Histoire forestière de l'Indochine (1850–1954) perspectives de recherché" [Forestry history of Indochina, 1850–1954: Research perspectives]. *Revue française d'histoire d'outre-mer* [French Review of Overseas History], 80: 219–50.

Bùi Dũng Thể, Đặng Thanh Hà, and Nguyễn Quốc Chính. 2004. *Rewarding Upland Farmers for Environmental Services: Experience, Constraints and Potential in Vietnam.* Bogor: World Agroforestry Centre (ICRAF) Southeast Asia Regional Office.

Bumpus, Adam. 2011. "The Matter of Carbon: Understanding the Materiality of tCO2e in Carbon Offsets." *Antipode* 43 (3): 612–38.

Burgess, Jacquelin, Judy Clark and Carolyn Harrison. 2000. "Knowledges in Action: An Actor Network Analysis of a Wetland Agri-Environment Scheme." *Ecological Economics* 35 (1): 119–32.

Byrne, Niall. 2008. "Rebuilding Vietnam's War-Torn Forests." *Ecos* 144: 26–28.

Callon, Michel. 1986. "Some Elements of a Sociology of Translation: Domestication of the Scallops and the Fishermen of St. Brieuc Bay." In *Power, Action and Belief*, edited by John Law, 196–223. London: Routledge.

———. ed. 1998. *The Laws of the Markets.* Oxford & Malden, MA : Blackwell Publishers.

———. 2009. "Civilizing Markets: Carbon Trading between In Vitro and In Vivo Experiments." *Accounting* 34: 535–48.

Callon, Michel, and Fabian Muniesa. 2005. "Peripheral Vision: Economic Markets as Calculative Collective Devices." *Organization Studies* 26 (8): 1229–50.

Cao Xuân Đức, Lưu Đức Xứng, and Trần Xán. 1965 [1882]. *Đại Nam nhất thống chí* [The Records of Unified Đại Nam]. Quyển thứ 13, *Tỉnh Hà Tĩnh* [Volume 13, Ha Tinh Province]. Saigon: Bộ Văn hoá Giáo dục [Ministry of culture and education].

Carolan, Michael. 2004. "Ontological Politics: Mapping a Complex Environmental Problem." *Environmental Values* 13 (4): 497–522.

Castella, Jean-Christophe, Stanislas Boissau, Nguyen Hai Thanh, and Paul Novosad. 2006. "Impact of Forestland Allocation on Land Use in a Mountainous Province of Vietnam." *Land Use Policy* 23 (2): 147–60.

Cederlöf, Gunnel, and K. Sivaramakrishnan, eds. 2006. *Ecological Nationalisms: Nature, Livelihoods, and Identities in South Asia.* Seattle: University of Washington Press.

Cepek, Michael. 2011. "Foucault in the Forest: Questioning Environmentality in Amazonia." *American Ethnologist* 38 (3): 501–15.

Cernea, Michael. 1992. "A Sociological Framework: Policy, Environment, and the Social Actors for Tree Planting." In *Managing the World's Forests: Looking for Balance Between Conservation and Development,* edited by Narendra Sharma, 301–35. Dubuque, IA: Kendall/Hunt Publishing.

Chomitz, Kenneth, Esteban Brenes, and Luis Constantino. 1999. "Financing Environmental Services: The Costa Rican Experience and its Implications." *Science of the Total Environment* 240: 157–69.

Chu Văn Tấn. 1962. *Một biến đổi cách mạng to lớn ở miền núi* [A large revolutionary change in the mountains]. Hanoi: NXB Sự thật [Truth publishing house].

Cleary, Mark. 2005a. "Managing the Forest in Colonial Indochina c. 1900–1940." *Modern Asian Studies* 39 (2): 257–83.

———. 2005b. "'Valuing the Tropics': Discourses of Development in the Farm and Forest Sectors of French Indochina, circa 1900–40." *Singapore Journal of Tropical Geography* 26 (3): 359–74.

Clement, Floriane, and Jamie Amezaga. 2008. "Linking Reforestation Policies with Land Use Change in Northern Vietnam: Why Local Factors Matter." *Geoforum* 39 (1): 265–77.

Cohen, Shaul. 1999. "Promoting Eden: Tree Planting as the Environmental Panacea." *Cultural Geographies* 6 (4): 424–46.

Cohen, Alice. 2012. "Rescaling Environmental Governance: Watersheds as Boundary Objects at the Intersection of Science, Neoliberalism, and Participation." *Environment and Planning A* 44 (9): 2207–24.

Conklin, Alice. 1997. *A Mission to Civilize: The Republican Idea of Empire in France and West Africa, 1895–1930*. Stanford: Stanford University Press.

Corbera, Esteve, Manuel Estrada, and Katrina Brown. 2010. "Reducing Greenhouse Gas Emissions from Deforestation and Forest Degradation in Developing Countries: Revisiting the Assumptions." *Climatic Change* 100 (3): 355–88.

Corbridge, Stuart, Glyn Williams, Manoj Srivastava, and Rene Veron. 2005. *Seeing the State: Governance and Governmentality in India*. Cambridge: Cambridge University Press.

Cư Hòa Vần. 1992. *The Development of the Ethnic Minorities of Vietnam*. Hanoi: Office for Ethnic Minorities and Mountainous Areas.

Cục Kiểm lâm. 1998. *Bản tin Kiểm lâm* [News from the Forest Protection Department]. Hanoi: Cục Kiểm lâm [Department of forest protection].

Đăng Minh. 2015. "Việc chặt hạ 6.700 cây xanh vi phạm Nghị định của Chính phủ" [Cutting 6,700 trees violates government decision]. *Công An Nhân Dân* [The people's police] March 22. http://kbchn.com/viec-chat-ha-6700-cay-xanh-vi-pham-nghi-dinh-cua-chinh-phu-11729.html.

Darier, Eric, ed. 1999. *Discourses of the Environment*. Oxford: Wiley-Blackwell.

Dauvergne, Peter. 1998. "Globalisation and Deforestation in the Asia-Pacific." *Environmental Politics* 7 (4): 114–35.

Davis, Diana. 2004. "Desert 'Wastes' of the Maghreb: Desertification Narratives in French Colonial Environmental History of North Africa." *Cultural Geographies* 11 (4): 359–87.

———. 2007. *Resurrecting the Granary of Rome: Environmental History and French Colonial Expansion in North Africa*. Athens, OH: Ohio University Press.

de Jong, Wil. 2009. "Forest Rehabilitation and Its Implication for Forest Transition Theory." *Biotropica* 42 (1): 3–9.

De Koninck, Rodolphe. 1999. *Deforestation in Viet Nam*. Ottawa: International Development Research Centre, Canada.

———. 2000. "The Theory and Practice of Frontier Development: Vietnam's Contribution." *Asia Pacific Viewpoint* 41 (1): 7–21.

———. 2004. "The Peasantry as the Territorial Spearhead of the State in Southeast Asia: The Case of Vietnam." *Sojourn* 11 (2): 231–58.

De Koninck, Rodolphe and Steve Déry. 1997. "Agricultural Expansion as a Tool of Population Redistribution in Southeast Asia." *Journal of Southeast Asian Studies* 28 (1): 1–26.

Dean, Mitchell. 1999. *Governmentality: Power and Rule in Modern Society*. Thousand Oaks, CA: Sage Publications .

Demeritt, Daniel. 2001. "Scientific Forest Conservation and the Statistical Picturing of Nature's Limits in the Progressive-era United States." *Environment and Planning D: Society and Space* 19: 431–59.

Department of Forest Development. 1998. *Proceedings of the National Seminar on Sustainable Forest Management and Forest Certification*. Hanoi: Department of Forest Development.

Déry, Steve. 2003. "Distinctions ethniques et déforestation au Việt Nam" [Ethnic distinctions and deforestation in Vietnam]. *Études rurales* [Rural studies], nos. 165–66, pp. 81–102.

Desbarats, Jacqueline. 1987. "Population Redistribution in the Socialist Republic of Vietnam." *Population and Development Review* 13 (1): 43–76.

Đinh Trần Dương. 2000. *Nghệ-Tĩnh với phong trào cách mạng giải phóng dân tộc trong 30 năm đầu thế kỷ XX* [Nghe Tinh and the revolutionary movements in the first 30 years of the 20th century]. Hà Nôi: NXB Chính trị [Politics publishing house].

Dove, Michael. 1983. "Theories of Swidden Agriculture, and the Political Economy of ignorance." *Agroforestry Systems* 29 (2): 337–51.

———. 2004. "Anthropogenic Grasslands in Southeast Asia: Sociology of Knowledge and Implications for Agroforestry." *Agroforestry Systems* 61: 423–35.

DPA. 2003. "Anti-smuggling Chief Caught Smuggling Listed Timber in Vietnam." April 16. Hanoi: Deutsche Press-Agentur.

Ducamp, Roger. 1912a. "L'envasement du port d'Haiphong et le régime forestier au Tonkin" [Siltation of Haiphong Harbor and the forest regime in Tonkin]. *Bulletin économique de l'Indochine* [Economic bulletin of Indochina], nouvelle séries, année 15, nos. 94–99, pp. 492–502.

———. 1912b. "La deforestation et la regime des pluies, profondeur des course d'eau, navigation" [Deforestation and the regime of rainfall, depth of water courses, and navigation] *Bulletin économique de l'Indochine* [Economic bulletin of Indochina], nouvelle séries, année 15, nos. 94–99, pp. 376–78.

Durst, Patrick, Thomas Waggener, Thomas Enters, and Tan Lay Cheng. 2001. *Forests Out of Bounds: Impacts and Effectiveness of Logging Bans in Asia-Pacific*. Bangkok: Asia-Pacific Forestry Commission and FAO Regional Office.

ESCAP. 1991. *Conservation and Management of Intertidal Forests in Viet Nam*. Bangkok: UN Economic and Social Commission for Asia and the Pacific.

Escobar, Arturo. 1995. *Encountering Development: The Making and Unmaking of the Third World*. Princeton, NJ: Princeton University Press.

———. 1999. "After Nature: Steps to an Antiessentialist Political Ecology." *Current Anthropology* 40 (1): 1–30.

Fairhead, James, and Melissa Leach. 1995. "False Forest History, Complicit Social Analysis: Rethinking Some West African Environmental Narratives." *World Development* 23 (6): 1023–35.

———. 1996. *Misreading the African Landscape: Society and Ecology in a Forest-Savanna Mosaic*. Cambridge, UK: Cambridge University Press.

———, eds. 1998. *Reframing Deforestation: Global Analyses and Local Realities: Studies in West Africa*. London: Routledge.

Fangeaux, G.D. 1931. "Le service forestier de l'Annam" [The forest service of Annam]. *Bulletin des amis du vieux Huế* [Bulletin of the friends of old Hue] 18: 228–37.

FAO. 2000. *Forest Resources Assessment 2000: On Definitions of Forest and Forest Change.* Rome: Food and Agriculture Organization of the United Nations.

———. 2005. *Global Forest Resources Assessment 2005.* Rome: Food and Agriculture Organization of the United Nations.

———. 2010. *Global Forest Resources Assessment 2010.* Rome: UN Food and Agriculture Organization of the United Nations.

Ferguson, James. 1994. *The Anti-Politics Machine: Development, Depoliticization, and Bureaucratic Power in Lesotho.* Minneapolis: University of Minnesota Press.

Fisher, Hayden, and Jenny Gordon. 2007. *Improved Australian Tree Species for Vietnam.* Impact Assessment Series No. 47. Canberra: Australian Centre for International Agricultural Research.

Ford, Carolyn. 2004. "Nature, Culture and Conservation in France and her Colonies, 1840–1940." *Past and Present* 183: 173–98.

Forsyth, Timothy. 1996. "Science, Myth and Knowledge: Testing Himalayan Environmental Degradation in Thailand." *Geoforum* 27 (3): 375–92.

———. 2002. *Critical Political Ecology: The Politics of Environmental Science.* London: Routledge.

———. 2005. "The Political Ecology of the Ecosystem Approach for Forests". In *Forests in Landscapes: Ecosystem Approaches for Sustainability,* edited by Jeffrey Sayer and Stewart Maginnis, 165–76. London: Earthscan.

———. 2011. "Politicizing Environmental Explanations: What Can Political Ecology Learn from Sociology and Philosophy of Science?" In *Knowing Nature: Conversations at the Intersection of Political Ecology and Science Studies,* edited by Mara Goldman, Paul Nadasdy, and Matthew Turner, 31–46. Chicago: University of Chicago Press.

Forsyth, Timothy, and Andrew Walker. 2008. *Forest Guardians, Forest Destroyers: The Politics of Environmental Knowledge in Northern Thailand.* Seattle: University of Washington Press.

Fortunel, Frédéric. 2000. *Le café au Viêtnam: de la colonisation à l'essor d'un grand producteur mondial* [Coffee in Vietnam: From colonization to the growth of a large global producer]. Paris: L'Harmattan.

———. 2009. "From Collectivization to Poverty, Natives and State-Owned Enterprises in Vietnam's Central Highlands. In *In the Name of Development: Indigenous Populations in South-East Asia under Command,* edited by Frédéric Boudier, 100–20. Bangkok: White Lotus Press.

Foucault, Michel. 2007. *Security, Territory, Population.* Translated by Graham Burchell. New York: Picador USA.

———. 2008. *The Birth Of Biopolitics: Lectures At The College De France, 1978–1979.* Translated by Graham Burchell. New York: Palgrave McMillian.

Foucault, Michel, Graham Burchell, Colin Gordon, and Peter Miller, eds. 1991. *The Foucault Effect: Studies in Governmentality.* Chicago: University of Chicago Press.

Fox, Jefferson, John Vogler, Omer Sen, Thomas Giambelluca, and Alan Ziegler. 2012. "Simulating Land-Cover Change in Montane Mainland Southeast Asia." *Environmental Management* 49 (5): 968–79.

Fox, Jefferson, Jean-Christophe Castella, and Alan Ziegler. 2014. "Swidden, Rubber and Carbon: Can REDD+ Work for People and the Environment in Montane Mainland Southeast Asia?" *Global Environmental Change Part A* 29: 318–26.

Galudra, G., and M. Sirait. 2009. "A Discourse on Dutch Colonial Forest Policy and Science in Indonesia at the Beginning of the 20th Century." *International Forestry Review* 11: 524–33.

Gayfer, Julian and Edwin Shanks. 1991. *Northern Vietnam: Farmers, Collectives and the Rehabilitation of Recently Reallocated Hill Land*. Network Paper 12a. London: Overseas Development Institute Social Forestry Network.

Gerber, Julien. 2011. "Conflicts over Industrial Tree Plantations in the South: Who, How and Why?" *Global Environmental Change* 21 (1): 165–76.

GGI (Gouvernement général de l'Indochine). 1933. *La terreur rouge en Annam (1930–1931): contribution a l'histoire des mouvements politiques de l'Indochine française* [The Red Terror in Annam 1930-31: Contribution to a history of political movements in French Indochina]. Hanoi: Gouvernement général de l'Indochine, Direction des affaires politiques et de la Sûreté Générale.

Glacken, Clarence. 1976. *Traces on the Rhodian Shore: Nature and Culture in Western Thought from Ancient Times to the End of the Eighteenth Century*. Berkeley: University of California Press.

Global Witness. 1999. *Made in Vietnam–Cut in Cambodia*. Phnom Penh: Global Witness.

———. 2000. *Chainsaws Speak Louder Than Words*. London: Global Witness.

Goldman, Michael. 2001. "Constructing an Environmental State: Eco-Governmentality and Other Transnational Practices of a 'Green' World Bank." *Social Problems* 48 (4): 499–523.

Gourou, Pierre. 1940. *Land Utilization in French Indochina*. Washington, DC: Institute of Pacific Relations.

Graham, Hamish. 1999. "Greedy or Needy? Forest Administration and Landowners' Attitudes in South-Western France during the Eighteenth Century." *Rural History* 16 (1): 1–20.

GSO. 1992. *Statistical Data of Vietnam's Agriculture, Forestry and Fishery, 1976–1991*. Hanoi: General Statistical Office.

Guha, Ramachandra. 1990. *The Unquiet Woods: Ecological Change and Peasant Resistance in the Himalaya*. Berkeley: University of California Press.

Guibier, Henri. 1924. *Note sur les reboisements* [Note on reforestation]. Hanoi: Imprimerie d'Extrême-Orient.

Guibier, J. 1918. *Situation des forêts de l'Annam* [Situation of the forests of Annam]. Saigon: Imprimerie Nouvelle Albert Portail.

Guillard, J. 2010. "Au service des forêts tropicales: Histoire des Services forestiers français outre-mer" [In the service of tropical forests: History of French Overseas forest services]. PhD thesis, L'École nationale du génie rural, des eaux et des forêts, Paris.

Hà Chu Chữ. 1995. *Forestry and Implementation of UNFCCC in Vietnam*. Hanoi: Forest Science Institute of Vietnam.

Hajer, Maarten. 1993. "Discourse Coalitions and the Institutionalization of Practice: The Case of Acid Rain in Britain." In *The Argumentative Turn in Policy Analysis and Plan-*

ning, edited by Frank Fisher and John Forester, 43–76. Durham, NC: Duke University Press.

Hall, Derek, Philip Hirsch, and Tania Li. 2011. *Powers of Exclusion: Land Dilemmas in Southeast Asia*. Honolulu: University of Hawai'i Press.

Hansen, M, P. Potapov, R. Moore, M. Hancher, S. Turubanova, A. Tyukavina et al. 2013. "High-Resolution Global Maps of 21st-Century Forest Cover Change." *Science* 342 (6160): 850–53.

Haraway, Donna. 1991. *Simians, Cyborgs, and Women: The Reinvention of Nature*. New York: Routledge.

Hardy, Andrew. 2005. *Red Hills: Migration and the State in the Highlands of Vietnam*. Copenhagen: Nordic Institute for Asian Studies Press.

Harrell, Stevan. 1995. "Introduction: Civilizing Projects and the Reaction to Them." In *Cultural Encounters on China's Ethnic Frontiers*, edited by Stevan Harrell, 3–36. Seattle: University of Washington Press.

Harzmann, Lutz. 1964. "Probleme in der Entwicklung der Forst -un Holzwirtschaft in der Demokratischen Republik Vietnam" [Problems in the development of the forest and wood industries in the Democratic Republic of Vietnam]. *Archiv fur Forstwesen* [Archive for forestry] 13: 1045–55.

Hay, Alistair. 1983. "Ho Chi Minh City Conference: Defoliants in Vietnam: The Long-Term Effects." *Nature* 302 (5905): 208–209.

Hickey, Gerald. 1982. *Sons of the Mountains: Ethnohistory of the Vietnamese Central Highlands to 1954*. New Haven: Yale University Press.

Hines, Deborah. 1995. *Financial Viability of Smallholder Reforestation in Vietnam*. Hanoi: Ministry of Forestry and United Nations Development Program.

Hoàng Cầm. 2007. *On Being "Forest Thieves": State Resource Policies, Market Forces and Struggles over Livelihood and Meaning of Nature in a Northwestern Valley of Vietnam*. Chiang Mai, Thailand: Regional Center for Social Science and Sustainable Development (RCSD).

Hoàng Cao Khải. 1915. "Les inondations au Tonkin [Floods in Tonkin]". *Bulletin économique de l'Indochine* [Economic bulletin of Indochina] nouvelle séries, 114 (July–August): 1–28.

Hoàng Hòe. 1996. *Deforestation and Reforestation in Vietnam*. Hanoi: Vietnam Forestry Science and Technical Association.

Hoàng Minh Hà, Đỗ Trọng Hoàn, Meine van Noordwijk, Phạm Thị Thu Thủy, Matilda Palm, Tô Xuân Phúc, Đoàn Diễm, Nguyễn Thanh Xuân, and Hoàng Thị Vân Anh. 2010. *An Assessment of Opportunities For Reducing Emissions From All Land Uses: Vietnam Preparing For REDD Final National Report*. Hanoi: Partnership for the Tropical Forest Margins.

Hoàng Xuân Ty. 1995. "Highland Development and Forest Resource Protection in Vietnam: Status and Research Priorities." In *The Challenges of Highland Development in Vietnam*, edited by Terry Rambo, Richard Reed, Lê Trọng Cúc, and Michael DiGregorio, 121–30. Honolulu: East-West Center.

H.T. 1980. "The Nghe Tinh Soviets," in "Nghe Tinh, Native Province of Ho Chi Minh," edited by Nguyễn Khắc Viện. *Vietnamese Studies* 59: 64–97.

Humphreys, David. 2008. *Logjam: Deforestation and the Crisis of Global Governance.* London: Routledge.

Hùng Tráng and Trọng Chàm. 2001. "Lạng Sơn phát triển kinh tế vườn-đồi-rừng" [Lang Son develops the garden-hill-forest economy]. *Nhân Dân* [The People], March 10, p. 2.

Hùng Võ. 2013. "Đổi mới nông-lâm trường: Người dân vẫn 'khát' đất [Renovating state agro-forestry enterprises: People are still 'thirsty' for land]." *Tiền Phong* [Pioneer], December 24. http://www.vietnamplus.vn/doi-moi-nonglam-truong-nguoi-dan-van-khat-dat/236519.vnp

ICARD. 2002. *The Impact of the Global Coffee Trade on Dak Lak Province, Viet Nam.* Hanoi: Information Centre for Agricultural and Rural Development and Oxfam.

Ingold, Timothy. 2000. *The Perception of the Environment: Essays on Livelihood, Dwelling and Skill.* London: Routledge.

Jackson, Larry. 1969. "The Vietnamese Revolution and the Montagnards." *Asian Survey* 9 (5): 313–30.

Jamieson, Neil. 1991. *Culture and Development in Vietnam.* Honolulu: East West Center.

Jarosz, Lucy. 2000. "Understanding Agri-Food Networks as Social Relations." *Agriculture and Human Values* 17 (3): 279–83.

Jasanoff, Sheila, ed. 2004. *States of Knowledge: The Co-Production of Science and Social Order.* New York: Routledge.

Jennings, Eric. 2004. *Vichy in the Tropics: Pétain's National Revolution in Madagascar, Guadeloupe, and Indochina, 1940–1944.* Stanford: Stanford University Press.

Jones, Owain, and Paul Cloke. 2002. *Tree Cultures: The Place of Trees and Trees in Their Place.* Oxford: Berg.

———. 2008. "Non-Human Agencies: Trees in Place and Time." In *Material Agency: Towards a Non-Anthropocentric Approach,* edited by Carl Knappett and Lambros Malafouris, 79–96. Dordrecht: Springer.

Kahin, George. 1972. "Minorities in the Democratic Republic of Vietnam." *Asian Survey* 11: 580–86.

Kemf, E. 1988. "The Re-Greening of Vietnam." *New Scientist* 1618: 53–57.

Kerkvliet, Benedict J.T. 1995. "Village-State Relations in Vietnam: The Effect of Everyday Politics on Decollectivization." *The Journal of Asian Studies* 54 (2): 396–418.

———. 2005. *The Power of Everyday Politics: How Vietnamese Peasants Transformed National Policy.* Ithaca, NY: Cornell University Press.

Kernan, Henry 1968. *Preliminary Report on Forestry in Vietnam.* Joint Development Group Working Paper No. 17. Saigon: USAID.

Kibel, Paul. 1995. "Legal Reform and the Fate of the Forests." *Environmental Policy and Law* 25: 1–6.

Kies, C.H.M. 1936. *Nature Protection in the Netherlands Indies.* Special Publication of the American Committee for International Wildlife Protection No. 8. Cambridge MA: American Committee for International Wildlife Protection.

Kim, Do-Hyung, Joseph Sexton, and John Townshend. 2015. "Accelerated Deforestation in the Humid Tropics from the 1990s to the 2000s." *Geophysical Research Letters* 42 (9): 3495–501.

Kirksey, Eban and Stefan Helmreich. 2010. "The Emergence of Multispecies Ethnography." *Cultural Anthropology* 25 (4): 545–76.

Kitchin, Rob and Martin Dodge. 2007. "Rethinking Maps." *Progress in Human Geography* 31 (3): 331–44.

Kleinen, John. 2011. "The Tragedy of the Margins: Land Rights and Marginal Lands in Vietnam (c. 1800–1945)." *Journal of the Economic and Social History of the Orient* 54 (4): 455–77.

Kosoy, Nicolas and Esteve Corbera. 2010. "Payments for Ecosystem Services as Commodity Fetishism." *Ecological Economics* 69 (6): 1228–36.

Kull, Christian, Charlie Shackleton, Peter Cunningham, Catherine Ducatillon, Jean-Marc Dufour-Dror, Karen Esler et al. 2011. "Adoption, Use and Perception of Australian Acacias around the World." *Diversity and Distributions* 17 (5): 822–36.

Kull, Christian, Xavier de Sartre, and Monica Castro-Larrañaga. 2015. "The Political Ecology of Ecosystem Services." *Geoforum* 61: 122–34.

Kummer, Daniel. 1992. *Deforestation in the Postwar Philippines*. Chicago: University of Chicago Press.

Lamb, David. 2011. *Regreening the Bare Hills: Tropical Forest Restoration in the Asia-Pacific Region*. Dordrecht: Springer.

Lambin, Eric F. and Patrick Meyfroidt. 2010. "Land use transitions: Socio-ecological feedback versus socio-economic change." *Land Use Policy* 27 (2), 108–18.

Lang, Chris. 2001. "Deforestation in Vietnam, Laos and Cambodia." In *Deforestation, Environment, and Sustainable Development: A Comparative Analysis*, edited by D. K. Vajpeyi, 111–37. New York: Praeger Publishers.

———. 2002. *The Pulp Invasion: The International Pulp and Paper Industry in the Mekong Region*. Montevideo, Uruguay: World Rainforest Movement.

Lankester, C.J. 1977. *War Damage, Problems and Perspectives for the Development of Forestry and Forest Industries in the Socialist Republic of Vietnam*. New York: Mission of the UN Secretary General.

Lansing, David. 2011. "Realizing Carbon's Value: Discourse and Calculation in the Production of Carbon Forestry Offsets in Costa Rica." *Antipode* 43 (3): 731–53.

Lastarria-Cornhiel, Susana. 1997. "Impact of Privatization on Gender and Property Rights in Africa." *World Development* 25: 1317–33.

Latour, Bruno. 1987. *Science in Action: How to Follow Scientists and Engineers Through Society*. Cambridge: Harvard University Press.

———. 1993. *We Have Never Been Modern*. Cambridge: Harvard University Press.

———. 2005. *Reassembling the Social: An Introduction to Actor-Network-Theory*. Oxford: Oxford University Press.

Laungaramsri, Pingkaew. 2000. "The Ambiguity of 'Watershed': The Politics of People and Conservation in Northern Thailand." *Sojourn: Journal of Social Issues in Southeast Asia* 15: 52–75.

Law, John. 1992. "Notes on the Theory of the Actor-Network: Ordering, Strategy, and Heterogeneity." *Systemic Practice and Action Research* 5 (4): 379–93.

Lê Đình Khả, Hà Huy Thịnh, and Nguyễn Việt Cường. 2003. "Improvement of Eucalypts for Reforestation in Vietnam." In *Eucalypts in Asia*, edited by J.W. Turnbull, 71–81. Canberra: Commonwealth Scientific and Industrial Research Organization.

Lê Hồng Tâm. 1972. "Về vấn đề xây dựng các vùng kinh tế mới" [On the problem of constructing new economic zones]. *Học Tập* [Academic Journal] 6: 48–59.

Lê Ngọc Thắng, Lê Hải Đường, Nguyễn Văn Thắng, Đô Văn Hoà, Nguyễn Lâm Thành, Hoàng Công Dũng, Hoàng Thị Lâm, Vũ Thanh Hiên, and Pamela McElwee. 2005. *Nghiên cứu về định cạnh, định cư ở Việt Nam* [Research on fixed cultivation and sedentarization in Vietnam]. Hanoi: NXB Chính trị Quốc gia [National political publishing house].

Lê Quý Đôn. 1977 [1776]. *Lê Quý Đôn toàn tập* [Collected works of Lê Quý Đôn], tập I: *Phủ biên tạp lục* [Volume 1: Miscellaneous chronicles of the pacified frontiers]. Hanoi: NXB Khoa học Xã hội [Social Sciences publishing house].

Lê Trọng Trải, Nguyễn Huy Dũng, Nguyễn Cử, Lê Văn Chẩm, Jonathan Eames, and Genevieve Chicoine. 2001. *An Investment Plan for Kẻ Gỗ Nature Reserve, Hà Tĩnh Province, Vietnam: A Contribution to the Management Plan*. Conservation Report Number 9. Hanoi: Birdlife International.

Lê Trung Đình. 1967. "Phát triển mạnh mẽ sản xuất nông nghiệp ở miền núi" [Developing strongly mountainous agricultural production]. *Học Tập* [Academic Journal] 1: 58–63.

Lecomte, Henri. 1923. *Les bois coloniaux* [The colonial woods]. Paris: Librarie Armand Colin.

Leimona, Beria, Grace Villamor, Meine van Noordwijk, Aunul Fauzi, and Retno Utaira. 2008. *Developing Mechanisms to Reward the Upland Poor in Asia for Environmental Services that They Provide*. Bogor, Indonesia: World Agroforestry Centre (ICRAF) Southeast Asia Regional Office.

Lentz, Christian. 2011. "Making the Northwest Vietnamese." *Journal of Vietnamese Studies* 6: 68–105.

Li, Tania. 2007. *The Will to Improve: Governmentality, Development, and the Practice of Politics*. Durham: Duke University Press.

Liljeström, Rita, Adam Fforde, and Bo Ohlsson. 1988. *Migrants by Necessity: A Report on the Living Conditions of Forestry Workers in the SIDA Supported Bai Bang Programme*. SIDA Evaluation Report, 1988/1. Stockholm: Swedish International Development Agency.

Liljeström, Rita, Eva Lindskog, Nguyễn Văn Áng, and Vương Xuân Tình. 1998. *Profit and Poverty in Rural Vietnam: Winners and Losers of a Dismantled Revolution*. Richmond, Surrey: Curzon.

Lindskog, Eva, Kirsten Dow, Goran Axberg, Fiona Miller, and Alan Hancock. 2005. *When Rapid Changes in Environmental, Social and Economic Conditions Converge: Challenges to Sustainable Livelihoods in Dak Lak, Vietnam*. Stockholm: Stockholm Environment Institute.

London, Jonathan. 2004. "Vietnam's Mass Education and Health Systems: A Regimes Perspective." *American Asian Review* 21: 125–70.

Long Hương. 1999. "Ngày thứ 3 xét xử vụ án phá rừng Tánh Linh [Day three of the trial of the case of deforestation in Tanh Linh]." *Lao Động* [Labor], April 1, p. 3.

Lorimer, Jamie. 2008. "Counting Corncrakes: The Affective Science of the UK Corncrake Census." *Social Studies of Science* 38 (3): 377–405.

Loschau, Manfred. 1963. "Zur Forstwirtschaft in der Demokratischen Republik Vietnam [On forestry in the Democratic Republic of Vietnam]." *Beiträge zur tropischen und subtropischen Landwirtschaft und Tropenveterinärmedizin* [Contributions to tropical and subtropical agriculture and veterinary medicine] 1 (1): 118–22.

———. 1969. "Zur wissenschaftlich-technischen Zusammenarbeit in der Forst- und Holzwirtschaft mit der Demokratischen Republic Vietnam [Scientific and technological cooperation with North Vietnam in the fields of forestry and forest products]." *Beiträge zur tropischen und subtropischen Landwirtschaft und Tropenveterinärmedizin* [Contributions to tropical and subtropical agriculture and veterinary medicine] 7 (3): 262–67.

Lövbrand, Eva and Johannes Stripple. 2011. "Making Climate Change Governable: Accounting for Carbon as Sinks, Credits and Personal Budgets." *Critical Policy Studies* 5: 187–200.

Lương, Hy Văn. 1992. *Revolution in the Village: Tradition and Transformation in North Vietnam, 1925–1988*. Honolulu: University of Hawai'i Press.

Lưu Hùng. 1986. "Forum: Vietnam's Central Highlands." *Vietnam Social Sciences* 1 (2): 146–68.

MacKenzie, Donald. 2009. "Making Things the Same: Gases, Emission Rights and the Politics of Carbon Markets." *Accounting, Organizations and Society* 34 (3–4): 440–55.

MacLean, Ken. 2013. *The Government of Mistrust: Illegibility and Bureaucratic Power in Socialist Vietnam*. Madison: University of Wisconsin Press.

Madec, J.H. 1997. "Retour sur le passé: La législation forestière tropical française" [Looking back: French tropical forest legislation." *Revue forestière française* [French forestry review] 49 (1): 69–78.

Malarney, Shaun. 2002. *Culture, Ritual and Revolution in Vietnam*. Honolulu: University of Hawai'i Press.

Mangin, Maurice. 1933. "Les forêts indochinoises, leur importance, leur gestion et leur mise en valeur" [The forests of Indochina: Their importance, management and development]". *Bulletin de la société d'encouragement pour l'industrie nationale* [Bulletin of the Society for the Encouragement of Domestic Industry] 132 (December): 641–63.

MARD. 1997. *Review Report on Planning, Organization and Management of Special Use Forest*. Hanoi: Ministry of Agriculture and Rural Development.

———. 2001a. *Five Million Hectare Reforestation Program Partnership Synthesis Report*. Hanoi: Ministry of Agriculture and Rural Development.

———. 2001b. *Lâm nghiệp Việt Nam 1945-2000: Quá trình phát triển và những bài học kinh nghiệm* [Forestry in Vietnam 1945-2000: Process of development and lessons learned]. Hanoi: NXB Nông nghiệp [Agriculture publishing house]

———. 2004. *Tải liệu Hội nghị tổng kết công tác định cảnh định cư giai đoạn 1990–2002* [Documents for the conference on summarizing the work of fixed cultivation and sedentarization from 1990–2002]. Hanoi: Ministry of Agriculture and Rural Development.

———. 2010a. *Decree 99 /2010/NĐ-CP On the Policy for Payment for Forest Environmental Services*. Hanoi: Ministry of Agriculture and Rural Development.

———. 2010b. *Diện tích rừng và đất lâm nghiệp toàn quốc: ban hành kèm theo Quyết định số 2140/QD-BNN-TCLN ngay 09/8/2010* [National area of forest and forestry land: Issued with Decision 2140]. Hanoi: Ministry of Agriculture and Rural Development.

———. 2011. *Báo cáo dự án trồng mới 5 triệu ha rừng giai đoạn 1998–2010* [Report on the five-million hectare reforestation project from 1998–2010]. Hanoi: Ministry of Agriculture and Rural Development.

———. 2014. *Forest Carbon Partnership Facility (FCPF) Carbon Fund Emission Reduc-*

tions Program Idea Note (ER-PIN): Sustainable Management of Forests Through Low Emissions Development Planning for Green Growth in the North Central Agro-Ecological Region. Hanoi: Ministry of Agriculture and Rural Development.

Marr, David. 1995. *Vietnam 1945: The Quest for Power*. Berkeley: University of California Press.

Martini, Edwin. 2012. *Agent Orange: History, Science, and the Politics of Uncertainty*. Amherst: University of Massachusetts Press.

Mather, Alexander. 2007. "Recent Asian Forest Transitions in Relation to Forest Transition Theory." *International Forestry Review* 9 (1): 491–502.

Mathews, Andrew. 2005. "Power/Knowledge, Power/Ignorance: Forest Fires and the State in Mexico." *Human Ecology* 33 (6): 795–820.

———. 2011. *Instituting Nature: Authority, Expertise, and Power in Mexican Forests*. Cambridge: MIT Press.

Maurand, Paul. 1943. *L'Indochine forestière* [Forestry in Indochina]. Hanoi: Imprimerie d'Extrême-Orient.

———. 1968. *Politique forestière à envisager au Viet Nam dans l'apres-guerre* [Forest policy for consideration after the war in Vietnam]. Saigon: Institut de Recherches Agronomiques.

McCarthy, John. 2002. "Power and Interest on Sumatra's Rainforest Frontier: Clientelist Coalitions, Illegal Logging and Conservation in the Alas Valley." *Journal of Southeast Asian Studies* 33 (1): 77–106.

MacDonald, Kenneth, and Catherine Corson. 2012. "'TEEB Begins Now': A Virtual Moment in the Production of Natural Capital." *Development and Change* 43 (1): 159–84.

McElwee, Pamela. 2002. "Lost Worlds and Local People: Protected Areas Development in Vietnam." In *Conservation and Mobile Indigenous Peoples: Displacement, Forced Settlement, and Sustainable Development*, edited by Dawn Chatty and Marcus Colchester, 312–29. Oxford: Berghahn Books.

———. 2004. "Becoming Socialist or Becoming Kinh? Government Policies for Ethnic Minorities in the Socialist Republic of Viet Nam." In *Civilizing the Margins: Southeast Asian Government Policies for the Development of Minorities*, edited by Christopher Duncan, 182–213. Ithaca: Cornell University Press.

———. 2005. "You Say Illegal, I Say Legal: The Relationship Between 'Illegal' Logging and Land Tenure, Poverty, and Forest Use Rights in Vietnam." *Journal of Sustainable Forestry* 19 (1–3): 97–135.

———. 2008. "Forest Environmental Income in Vietnam: Household Socioeconomic Factors Influencing Forest Use." *Environmental Conservation* 35 (2): 147–59.

———. 2009. "Reforesting 'Bare Hills' in Vietnam: Social and Environmental Consequences of the 5 Million Hectare Reforestation Program." *AMBIO: a Journal of the Human Environment* 38 (6): 325–33.

———. 2010. "Resource Use among Rural Agricultural Households near Protected Areas in Vietnam: The Social Costs of Conservation and Implications for Enforcement." *Environmental Management* 45 (1): 113–31.

———. 2012. "Payments for Environmental Services as Neoliberal Market-based Forest Conservation in Vietnam: Panacea or Problem?" *Geoforum* 43 (3): 412–26.

McElwee, Pamela, Tuyen Nghiem, Hue Le, Huong Vu, and Nghi Tran. 2014. "Payments for

Environmental Services and Contested Neoliberalisation in Developing Countries: A Case Study from Vietnam." *Journal of Rural Studies* 36: 423–40.

McElwee, Pamela, and Nguyễn Chí Thanh 2015. *Report on 3 Years of Implementation of the Payments for Environmental Services Policy.* Hanoi: Winrock International, USAID, and VNFOREST.

MEA [Millennium Ecosystem Assessment]. 2005. *Millennium Ecosystem Assessment Synthesis Report.* Washington, DC: World Resources Institute.

Mellac, Marie. 1997. "L'etat et la forêt au nord Vietnam" [The state and the forest in North Vietnam]." *Cahiers d'outre-mer* [Overseas notebooks] 50 (197): 27–42.

———. 1998. "La gestion des espaces forestiers au Nord Vietnam: Un modèle et son application [The management of forest areas in North Vietnam: A model and its applications]." *Cahiers d'outre-mer* [Overseas notebooks] 51 (204): 339–66.

Merrill, Elmer. 1933. "Loureiro and his Botanical Work." *Proceedings of the American Philosophical Society* 72 (4): 229–39.

Messier, Christian, Klaus Puettmann, and David Coates. 2013. *Managing Forests as Complex Adaptive Systems: Building Resilience to the Challenge of Global Change.* London: Earthscan from Routledge.

Meyfroidt, Patrick, and Eric Lambin. 2008a. "Forest Transition in Vietnam and Its Environmental Impacts." *Global Change Biology* 14 (6): 1319–36.

———. 2008b. "The Causes of the Reforestation in Vietnam." *Land Use Policy* 25 (2): 182–97.

———. 2009a. "Forest Transition in Vietnam and Bhutan: Causes and Environmental Impacts." In *Reforesting Landscapes: Linking Pattern and Process*, edited by Harini Nagendra and Jane Southworth, 315–339. Dordrecht: Springer.

———. 2009b. "Forest Transition in Vietnam and Displacement of Deforestation Abroad." *Proceedings of the National Academy of Sciences of the United States of America* 106 (38): 16139–44.

Meyfroidt, Patrick, Vũ Tấn Phương, and Hoàng Việt Anh. 2013. "Trajectories of Deforestation, Coffee Expansion and Displacement of Shifting Cultivation in the Central Highlands of Vietnam." *Global Environmental Change* 23 (5): 1187–98.

Miller, Peter and Nikolas Rose. 2008. *Governing the Present: Administering Economic, Social and Personal Life.* London: Polity Press.

Mitchell, Timothy. 1991. "The Limits of the State: Beyond Statist Approaches and their Critics." *The American Political Science Review* 85: 77–96.

———. 2002. *Rule of Experts: Egypt, Techno-politics, Modernity.* Berkeley: University of California Press.

Mitlöhner, Ralph, and Chaw Chaw Sein. 2011. *Acacia Hybrid: Ecology and Silviculture in Vietnam.* Bogor: Center for International Forestry Research (CIFOR).

MOF (Ministry of Forest). 1990. *Báo cáo phương án đổi mới công tác định canh định cư* [Report on renovation approaches to the work of fixed cultivation and settlement]. Hanoi: Ministry of Forestry.

———. 1991a. *30 năm xây dựng và phát triển ngành lâm nghiệp, 1961–1990* [30 years of constructing and developing the forestry sector, 1961–90]. Hanoi: Ministry of Forestry.

———. 1991b. *Vietnam Forestry Sector Review, Tropical Forestry Action Programme.* Hanoi: Ministry of Forestry.

———. 1995. Vietnam Forestry. Hanoi: Agricultural Publishing House.
Moise, Edwin. 1983. *Land Reform in China and North Vietnam: Consolidating the Revolution at the Village Level*. Chapel Hill: University of North Carolina Press.
Moore, Donald. 2000. "The Crucible of Cultural Politics: Reworking 'Development' in Zimbabwe's Eastern Highlands." *American Ethnologist* 26 (3): 654–89.
MoSTE [Ministry of Science, Technology and Environment]. 1996. *Proceedings of the Conference, 'Creating Revenues from Biodiversity in Order to Conserve It,' 21–23 November*. Hanoi: Ministry of Science, Technology and Environment and Danish International Development Agency.
Murdoch, Jonathan. 1997. "Inhuman/Nonhuman/Human: Actor-Network Theory and the Prospects for a Nondualistic and Symmetrical Perspective on Nature and Society." *Environment and Planning D: Society and Space* 15: 731–56.
Nambiar, Sadanandan, Christopher Harwood, and Nguyễn Đức Kiên. 2015. "Acacia Plantations in Vietnam: Research and Knowledge Application to Secure a Sustainable Future." *Southern Forests* 77 (1): 1–10.
Neitsch, S.L., J.G. Arnold and J.R. Williams. 2001. *Soil and Water Assessment Tool*. College Station, TX: Texas A&M University.
Nghiêm Thị Yến. 1994. "Report on Vietnamese Women Forestry Workers." Paper presented at the National Symposium on the Integration of Women in Agricultural and Rural Development in Vietnam, Hanoi. Hanoi: Food and Agriculture Organization.
Ngô Đình Quê, Nguyễn Thu Hương, Nguyễn Thanh Tùng, and Tạ Thu Hoà. 2006. *Đánh giá ảnh hưởng của cây công nghiệp thân gỗ (cà phê) đến môi trường ở Tây Nguyên* [An evaluation of the impact of industrial crops (coffee) on the environment in the Central Highlands]. Hanoi: Forest Science Institute of Vietnam.
Ngô Đình Thọ, Phạm Xuân Phương, Bùi Huy Nho, and Nguyễn Hữu Tuynh. 2006. *Cẩm nang ngành lâm nghiệp: Quản lý lâm trường quốc doanh* [Forestry review: Management of state forest enterprises]. Hanoi: Ministry of Agriculture and Rural Development.
Ngô Vĩnh Long. 1990. "Communal Property and Peasant Revolutionary Struggles in Vietnam." *Peasant Studies* 17: 121–40.
Nguyễn Cúc Sinh, and Phan Lâm Lương. 1986. *Thống kê lâm nghiệp* [Forest statistics]. Hanoi: NXB Thông kê [Statistics publishing house].
Nguyễn Huy Thiệp. 2003. *Crossing the River: Short Fiction*. Translated by Nguyễn Nguyệt Cẩm and Dana Sachs. Willimantic, CT: Curbstone Press.
Nguyễn Phúc Khánh, and Nguyễn Văn Trọng, eds. 1986. *Nghệ Tĩnh hôm qua và hôm nay* [Nghe Tinh yesterday and today]. Hanoi: NXB Sự thật [Truth publishing house].
Nguyễn Quang Tân. 2006. "Forest Devolution in Vietnam: Differentiation in Benefits from Forest among Local Households." *Forest Policy and Economics* 8 (4): 409–20.
———. 2011. *Payment for Environmental Services in Vietnam: An Analysis of the Pilot Project in Lam Dong Province*. Japan: Institute for Global Environmental Strategies.
Nguyễn Sinh Hùng. 2010. *Ý kiến kết luận của Phó Thủ tướng Nguyễn Sinh Hùng tại hội nghị sơ kết chính sách thí điệm chi trả dích vụ môi trường rừng, theo Quyét định số 380/QD-TTg ngày 10 tháng 4 năm 2008 của Thủ tướng Chính phủ* [Summary opinion of Vice Prime Minister Nguyen Sinh Hung at the conference to review the pilot policy on forest environmental services, according to Decision 380 of the government]. Hanoi: Văn phòng Chính phủ [State Government Office].

Nguyễn Tạo. 1968. *Rừng và nghề rừng: Phục vụ sự nghiệp công nghiệp hóa xã hội chủ nghĩa ở miền Bắc nước ta* [Forests and forestry: Serving the work of industrialization of socialism in the North of our country.] Hanoi: NXB Nông thôn [Rural publishing house].

Nguyễn Tiến Hưng. 1989. *Economic Development of Socialist Vietnam, 1955–80*. New York: Praeger Publishers.

Nguyễn Thị Thanh Hải. 1999. "Lâm tặc làm chủ rừng Yên Tú" [Illegal loggers are the owners of the forests of Yen Tu]. *Lảo Động* [Labor], July 24, p. 1.

Nguyễn Trọng Cồn. 1980. *Phong trào công nhân trong cao trào Xô viết Nghệ Tĩnh* [Movement of the workers in the grand Nghe Tinh Soviets uprising]. Hanoi: NXB Lao động [Labor publishing house].

Nguyễn Vành, Trần Mạnh Cương & Trần Đình Đàn. 1995. *Strategy for Regreening Barren Lands and Hills in Viet Nam*. Hanoi: Ministry of Agriculture and Rural Development.

Nguyễn Văn Chính. 2008. "From Swidden Cultivation to Fixed Farming and Settlement: Effects of Sedentarization Policies among the Kmhmu in Vietnam." *Journal of Vietnamese Studies* 3 (3): 44–80.

Nguyễn Văn Công. 2001. "Người biết phát huy phong trào Tết trồng cây của Bác Hồ" [Promoting the Tet tree planting movement of Uncle Ho.] *Nhân Dân* [The People], February 1, p. 2.

Nguyễn Văn Sản, and Don Gilmour. 1999. *Forest Rehabilitation Policy and Practice in Vietnam: Proceedings of a National Workshop*. Hanoi: International Union for the Conservation of Nature.

Nguyễn Xuân Khẳng. 1965. *Trồng rừng phòng hộ phục vụ thâm canh tăng năng suất cây trồng* [Planting protection forests to intensify and increase productivity of planted trees]. Hanoi: NXB Nông thôn [Agriculture publishing house].

Nikolic, Nina, Rainer Schultze-Kraft, Miroslave Nikolic, Reinhard Böcker, and Ingo Holz. 2008. "Land Degradation on Barren Hills: A Case Study in Northeast Vietnam." *Environmental Management* 42 (1): 19–36.

Nông Quốc Chấn. 1978. "Thirty Years of Cultural Work among the Ethnic Minorities." *Vietnamese Studies* 52: 57–63.

Oels, Angela. 2005. "Rendering Climate Change Governable: From Biopower to Advanced Liberal Government?" *Journal of Environmental Policy & Planning* 7 (3): 185–207.

Ogle, Alan, Jay Blakeney, and Hoàng Hòe. 1998. *Natural Forest Management Practices*. Evaluation of State Forest Enterprises Working Paper No. 2. Hanoi: Ministry of Agriculture and Rural Development and Asian Development Bank.

Ogonowski, Matthew and Adrian Enright. 2013. *Cost Implications for Pro-Poor REDD+ in Lam Dong Province, Vietnam: Opportunity Costs and Benefit Distribution Systems*. London: International Institute for Environment and Development.

Ohlsson, Bo, Mats Sandewall, Kajsa Sandewall, and Nguyễn Huy Phồn. 2005. "Government Plans and Farmers Intentions: A Study on Forest Land Use Planning in Vietnam." *AMBIO: A Journal of the Human Environment* 34 (3): 248–55.

Ortner, Sherry. 1995. "Resistance and the Problem of Ethnographic Refusal." *Comparative Studies in Society and History* 37 (1): 173–93.

Painter, Joe. 2010. "Rethinking Territory." *Antipode* 42 (5): 1090–118.

Paterson, Matthew, and Johannes Stripple. 2010. "My Space: Governing Individuals' Carbon Emissions." *Environment and Planning D: Society and Space* 28: 341–62.

Peet, Richard, Paul Robbins, and Michael Watts, eds. 2011. *Global Political Ecology*. New York: Routledge.

Pelley, Patricia. 2002. *Postcolonial Vietnam: New Histories of the National Past*. Durham, NC: Duke University Press.

Peluso, Nancy Lee. 1992. *Rich Forests, Poor People: Resource Control and Resistance in Java*. Berkeley: University of California Press.

———. 2003. "Territorializing Local Struggles for Resource Control: A Look at Environmental Discourses and Politics in Indonesia." In *Nature in the Global South: Environmental Projects in South and Southeast Asia*, edited by Paul Greenhough and Anna Tsing, 231–52. Durham: Duke University Press.

Peluso, Nancy Lee, and Peter Vandergeest. 2001. "Genealogies of the Political Forest and Customary Rights in Indonesia, Malaysia, and Thailand." *Journal of Asian Studies* 60: 761–812.

Peytavin, M. 1916. *Rapport sur la crue du Fleuve Rouge et les inondations du Tonkin en 1915* [Report on the flood on the Red River and the flooding in Tonkin 1915]. Hanoi: Gouvernement général de l'Indochine, Administration des travaux publics.

Phạm Hưng. 2000. "Khi lâm tặc được bồi thương: Hoàng Minh Huệ không phạm tội, 'làm chết người trong khi thi hành công vụ'" [When illegal loggers are compensated: Hoanh Minh Hue is not guilty, "Killing of a person was carried out in the course of fulfilling a duty"]. *Lao Động* [Labor], August 15, p. 1.

Phạm Hữu Văn. 1989. *Easup Forestry-Agricultural-Industry Union: Forest Management Plan and Results on 10 Years of the Implementation*. Dak Lak, Vietnam: Easup Forestry-Agricultural-Industry Union.

Phạm Thị Thu Thủy, Moira Moeliono, Nguyễn Thị Hiên, Nguyễn Hữu Thọ, and Vũ Thị Hiên. 2012. *The Context of REDD+ in Vietnam: Drivers, Agents and Institutions*. Bogor: Center for International Forestry Research.

Phạm Thị Thu Thủy, Karen Bennett, Vũ Tấn Phương, Jake Brunner, Lê Ngọc Dũng, and Nguyễn Đình Tiến. 2013. *Payments for Forest Environmental Services in Vietnam: From Policy to Practice*. Bogor: Center for International Forestry Research.

Phạm Thị Thu Thủy, Monica Di Gregorio, Rachel Carmenta, Maria Brockhaus and Lê Ngọc Dũng. 2014. "The REDD+ Policy Arena in Vietnam: Participation of Policy Actors." *Ecology and Society* 19 (2): art22.

Phạm Văn Duẩn and Phùng Văn Khoa. 2013. "Xây dựng bản đồ hệ số k phục vụ chi trả dịch vụ môi trường rừng trong lưu vực" [Mapping the K coefficient for the payment for forest environmental services in watersheds.] *Tạp chí Khoa học Lâm nghiệp* [Forestry science journal] 2: 2753–63.

Phan, John. 2010. "Re-imagining 'Annam': A New Analysis of Sino–Viet–Muong Linguistic Contact." *Chinese Southern Diaspora Studies* 4: 3–24.

Phan Hồng. 2006. "Hà Tĩnh: Chi Cục trưởng Kiểm lâm cấp thư tay cho chở gỗ . . . lậu! [Ha Tinh: The head of the forest ranger service provides handwritten letter to transport wood…illegally!]" *VietnamNet*, December 8. http://vietbao.vn/An-ninh-Phap-luat/Chi-Cuc-truong-Kiem-lam-cap-thu-tay-cho-cho-go . . . -lau!/20641332/218/

Phan Xuân Đợt. 1984. *Phát triển nghề rừng theo hướng sản xuất lớn xã hội chủ nghĩa* [Developing forestry according to large scale socialism]. Hanoi: NXB Sự thật [Truth publishing house].

Phinney, Harriet. 2005. "Asking for a Child: The Refashioning of Reproductive Space in Post-War Northern Vietnam." *The Asia Pacific Journal of Anthropology* 6 (3): 215–30.

Pierre, A., Clovis Thorel, and Francis Garnier. 1866. "Rapport de la commission chargée d'étudier les questions qui se rapportent au commerce du bois et à l'exploitation des forêts" [Report of the commission charged with studying the questions concerning wood commerce and forest exploitation]. *Bulletin du comité agricole et industriel de la Cochinchine* [Bulletin of the agricultural and industrial committee of Cochinchina] 1: 6–21.

Pincetl, Stephanie. 1993. "Some Origins of French Environmentalism: An Exploration." *Forest & Conservation History* 37 (2): 80–89.

Polanyi, Karl. 1944. *The Great Transformation*. New York: Beacon Press.

Prudham, Scott. 2003. "Taming Trees: Capital, Science, and Nature in Pacific Slope Tree Improvement." *Annals of the Association of American Geographers* 9 (3): 636–56.

———. 2004. *Knock on Wood: Nature as Commodity in Douglas Fir Country*. New York: Routledge.

Rambo, Terry, Keith Fahrney, Trần Đức Viên, Jeff Romm, and Dang Thi Sy. 1996. *Red Books, Green Hills: The Impact of Economic Reform on Restoration Ecology in the Midlands of Northern Vietnam*. Honolulu: East-West Center.

Rival, Laura, ed. 1998. *The Social Life of Trees: Anthropological Perspectives on Tree Symbolism*. London: Bloomsbury.

Robbins, Paul. 2001a. "Tracking Invasive Land Covers in India, or Why Our Landscapes Have Never Been Modern." *Annals of the Association of American Geographers* 91 (4): 637–59.

———. 2001b. "Fixed Categories in a Portable Landscape: The Causes and Consequences of Land Cover Categorization." *Environment and Planning A* 33: 161–79.

Roe, Emery. 1991. "Development Narratives, or Making the Best of Blueprint Development." *World Development* 19 (4): 287–300.

Rose, Nikolas. 1999. *Powers of Freedom: Reframing Political Thought*. Cambridge, UK: Cambridge University Press.

Rutherford, Stephanie. 2007. "Green Governmentality: Insights and Opportunities in the Study of Nature's Rule." *Progress in Human Geography* 31 (3): 291–307.

Sack, Robert. 1986. *Human Territoriality: Its Theory and History*. Cambridge, UK: Cambridge University Press.

Salemink, Oscar. 1997. "The King of Fire and Vietnamese Ethnic Policy in the Central Highlands." In *Development or Domestication? Indigenous Peoples of Southeast Asia*, edited by Don McCaskill and Ken Kampe, 488–535. Bangkok: Silkworm Books.

———. 2003. *The Ethnography of Vietnam's Central Highlanders: A Historical Contextualization, 1850–1990*. Honolulu: University of Hawai'i Press.

Sanchez-Azofeifa, Arturo, Alexander Pfaff, Juan Robalino, and Judson Boomhower. 2007. "Costa Rica's Payment for Environmental Services Program: Intention, Implementation, and Impact." *Conservation Biology* 21 (5): 1165–73.

Sandewall, Mats, Bo Ohlsson, Kajsa Sandewall, and Le Sy Viet. 2010. "The Expansion of Farm-based Plantation Forestry in Vietnam." *AMBIO: A Journal of the Human Environment* 39 (8): 567–79.

Sarraut, Albert. 1913. *Arrêté du Gouverneur Général concernant les bois et forêts sis sur le territoire de la Cochinchine* [Order of the Governor General regarding woods and forests located in the territory of Cochinchina]. Hanoi: Gouvernement général de l'Indochine .

Schroeder, Richard. 1997. "'Re-claiming' Land in The Gambia: Gendered Property Rights and Environmental Intervention." *Annals of the Association of American Geographers* 87 (3): 487–508.

———. 1999. *Shady Practices: Agroforestry and Gender Politics in The Gambia.* Berkeley: University of California Press.

Scott, James C. 1976. *The Moral Economy of the Peasant: Rebellion and Subsistence in Southeast Asia.* New Haven : Yale University Press.

———. 1985. *Weapons of the Weak: Everyday Forms of Peasant Resistance.* New Haven: Yale University Press.

———. 1990. *Domination and the Arts of Resistance: Hidden Transcripts.* New Haven: Yale University Press.

———. 1998. *Seeing Like a State: How Certain Schemes to Improve the Human Condition Have Failed.* New Haven: Yale University Press.

———. 2009. *The Art of Not Being Governed: An Anarchist History of Upland Southeast Asia.* New Haven: Yale University Press.

Scott, James C., and Benedict Kerkvliet, eds. 1986. *Everyday Forms of Peasant Resistance in Southeast Asia.* London: Frank Cass.

Service géographique de l'Indochine. 1925. *Carte économique de l'Annam, dressée par les Services agricoles* (1:100,000) [Economic map of Annam, prepared by the Agricultural Service]. Paris: Service géographique de l'Indochine.

———. 1925. *Carte économique de l'Annam Province de Ha Tinh* [Economic map of Annam, Province of Ha Tinh]. Paris: Service géographique de l'Indochine.

Service scientifique de l'Agence économique de l'Indochine. 1931. *La situation forestière de l'Indochine* [The forestry situation in Indochina]. Paris: Exposition Coloniale Internationale Paris.

Sikor, Thomas. 1995. "Decree 327 and the Restoration of Barren Land in the Vietnamese Highlands." In *The Challenges of Highland Development in Vietnam*, edited by Terry Rambo, Robert Reed, Lê Trọng Cúc, and Michael Digregorio, 143–56. Honolulu: East-West Center.

———. 2012. "Tree Plantations, Politics of Possession and the Absence of Land Grabs in Vietnam." *Journal of Peasant Studies* 39 (3–4): 1077–101.

Sikor, Thomas, and Đào Minh Trường. 2000. *Sticky Rice, Collective Fields: Community-Based Development among the Black Thai.* Hanoi: Agricultural publishing house.

Sikor, Thomas, and Nguyễn Quang Tân. 2007. "Why May Forest Devolution Not Benefit the Rural Poor? Forest Entitlements in Vietnam's Central Highlands." *World Development* 35 (11): 2010–25.

Sikor, Thomas, and Tô Xuân Phúc. 2011. "Illegal Logging in Vietnam: Lam Tac (Forest Hijackers) in Practice and Talk." *Society and Natural Resources* 24 (7): 688–701.

Sikor, Thomas, and Trần Ngọc Thanh. 2007. "Exclusive versus Inclusive Devolution in Forest Management: Insights from Forest Land Allocation in Vietnam's Central Highlands." *Land Use Policy* 24 (4): 644–53.

Sion, Jules. 1920. "Le déboisement et les inondations au Tonkin [Deforestation and flooding in Tonkin.]" *Annales de géographie* 29: 315–17.

Sivaramakrishnan, K. 1995. "Colonialism and Forestry in India: Imagining the Past in Present Politics." *Comparative Studies in Society and History* 37 (1): 3–40.

———. 1999. *Modern Forests: Statemaking and Environmental Change in Colonial Eastern India*. Stanford, CA: Stanford University Press.

———. 2000. "State Sciences and Development Histories: Encoding Local Forestry Knowledge in Bengal." *Development and Change* 31: 61–89.

Smith, G. A. 1993. *Working Paper: Livestock and Barren Land Development*. Hanoi: Food and Agriculture Organization of the United Nations (FAO).

Sowerwine, Jennifer. 2004. "Territorialisation and the Politics of Highland Landscapes in Vietnam: Negotiating Property Relations in Policy, Meaning and Practice." *Conservation and Society* 2 (1): 97–136.

SRV [Socialist Republic of Vietnam]. 1999. *Decision of the Government No. 187 QĐ-TTg on the Renovation of Organization and Mechanisms for Management of State Forest Enterprises*. Hanoi: Socialist Republic of Vietnam.

———. 2001. *Di dân, kinh tế mới, định canh định cư: lịch sự và truyền thống* [Migration, new economic zones, fixed cultivation and settlement: history and tradition]. Hanoi: Socialist Republic of Vietnam.

———. 2011. *Báo cáo: Tổng kết thực hiện Dự án "Trồng mới 5 triệu ha rừng" và kế hoạch bảo vệ, phát triển rừng giai đoạn 2011–2020* [Report: Summary of the implementation of the 5 million hectare reforestation project and plan for protecting and developing forests from 2011–20]. Hanoi: Office of the Prime Minister, Socialist Republic of Vietnam.

Stanton, Tracy, Marta Echavarria, Katherine Hamilton, and Caroline Ott. 2010. *State of Watershed Payments: An Emerging Marketplace*. Washington, DC: Ecosystem Marketplace.

Star, Susan. 2010. "This is Not a Boundary Object: Reflections on the Origin of a Concept." *Science, Technology & Human Values* 35 (5): 601–17.

Star, Susan, and James Griesemer. 1989. "Institutional Ecology, 'Translations' and Boundary Objects: Amateurs and Professionals in Berkeley's Museum of Vertebrate Zoology, 1907–39." *Social Studies of Science* 19 (3): 387–420.

Stellman, Jenne, Steven Stellman, Richard Christian, Tracy Weber, and Carrie Tomasallo. 2003. "The Extent and Patterns of Usage of Agent Orange and Other Herbicides in Vietnam." *Nature* 422 (6933): 681–87.

Swanson, Carl. 1975. "Reforestation in the Republic of Vietnam." *Journal of Forestry* 73 (6): 367–71.

Swyngedouw, Erik. 2004. *Social Power and the Urbanization of Water*. Oxford: Oxford University Press.

Tạ Văn Tài. 1985. "Ethnic Minorities and the Law in Traditional Vietnam." *Vietnam Forum* 5: 22-36.

Thái Văn Trừng. 1970. *Thẩm thực vật rừng Việt Nam* [Forest vegetation in Vietnam]. Hanoi: NXB Khoa học và kỹ thuật [Science and technology publishing house].

Thomas, Frédéric. 1998. "Ecologie et gestion forestiere dans l'Indochine française [Ecology and forest management in French Indochina]". *Revue française d'histoire d'outre-mer* [French journal of overseas history] 319: 59–86.

———. 1999. *Histoire du régime et des services forestiers française en Indochine de 1862 à 1945* [History of the regime and the French forest services in Indochina from 1862–1945]. Hanoi: Thế Giới [The world] publishers.

———. 2000. "Forêts de Cochinchine et 'bois coloniaux', 1862-1900" [Forests of Cochinchina and 'colonial woods']. *Autrepart* 15: 49–72.

———. 2009. "Protection des forêts et environmentalisme coloniale: Indochine 1860–1945 [Forest protection and colonial environmentalism: Indochina, 1860–1945]." *Revue d'histoire moderne & contemporaine* [Review of modern and contemporary history] 56: 104–36.

Thorel, Clovis. 2001 [1873]. *Agriculture and ethnobotany of the Mekong Basin [Voyage d'exploration en Indo-Chine Volume 2].* Translated by Walter Tips. Bangkok: White Lotus Press.

Thulstrup, Andreas. 2014. "Plantation Livelihoods in Central Vietnam: Implications for Household Vulnerability and Community Resilience." *Norsk Geografisk Tidsskrift–Norwegian Journal of Geography* 68 (1): 1–9.

Thulstrup, Andreas, Thorkil Casse, and Thomas Nielsen. 2013. "The Push for Plantations: Drivers, Rationales and Social Vulnerability in Quang Nam Province, Vietnam." In *On the Frontiers of Climate and Environmental Change: Vulnerabilities and Adaptations in Central Vietnam*, edited by Ole Bruun and Thorkil Casse, 71–89. Heidelberg: Springer.

Thúy Hạnh. 2015. "Hà Nội 'trồng nhầm' gỗ mỡ, không phải vàng tâm?" [Hanoi planting wrong magnolia trees?] *VietnamNet Online*, March 21. http://vietnamnet.vn/vn/xa-hoi/226920/ha-noi--trong-nham--go-mo--khong-phai-vang-tam-.html.

Tô Xuân Phúc, Wolfram Dressler, Sango Mahanty, Phạm Thị Thu Thủy, and Claudia Zingerli. 2012. "The Prospects for Payment for Ecosystem Services (PES) in Vietnam: A Look at Three Payment Schemes." *Human Ecology* 40 (2): 237–49.

Tô Xuân Phúc, Phan Đình Nhã, Phạm Quang Tú, and Đỗ Duy Khôi. 2013. *Mâu thuẫn đất đai giữa công ty lâm nghiệp và người dân địa phương* [Conflicts over land between state forest companies and local people]. Hanoi: Forest Trends.

Tô Xuân Phúc, and Trần Hữu Nghị. 2014. *Rubber Expansion and Forest Protection in Vietnam.* Hue, Viet Nam: Forest Trends and Tropenbos International.

Trần Huy Liệu. 1961. "Vấn đề chính quyền Xô Việt" [The problem of the authorities of the Soviets]. *Nghiên cứu lịch sử* [Historical research] 33 (12): 1–7.

Trần Lộc. 1971. "Yên Bái, tỉnh miền núi làm tốt công tác quân sự địa phương [Yen Bai, a mountainous province doing well in local military work.]" *Học Tập* [Academic Journal] 5: 43–49.

Trần Ngọc Thanh and Thomas Sikor. 2006. "From Legal Acts to Actual Powers: Devolution and Property Rights in the Central Highlands of Vietnam." *Forest Policy and Economics* 8 (4): 397–408.

Trần Xuân Thiệp. 1996. *Eucalyptus Plantations in Vietnam: Their History and Development Process.* Bangkok: FAO Regional Office for Asia and the Pacific.

Trường Chinh. 1983. "Đưa nhân dân các dân tộc ở Đăk Lăk tiến thẳng lên chủ nghĩa xã hội [Taking the people of the ethnic minority groups of Dak Lak to reach socialism]." *Tạp chí Cộng sản* [Communist journal] 8: 7–18.

Trung tâm Nghiên cứu Phụ nữ [Research Center for Women]. 1987. *Kỷ yếu hội thảo khoa học về điều kiện lao động và đời sống của nữ công nhân lâm trường nguyên liệu giấy tỉnh Hà Tuyên* [Proceedings of the scientific workshop on the conditions of laborers and lives of women workers in the state forest enterprises for paper production, Ha Tuyen province]. Hanoi: Research Center for Women.

UN-REDD. 2013. *Legal Analysis of Cross-Cutting Issues for REDD+ Implementation: Lessons Learned from Mexico, Viet Nam and Zambia*. Geneva: UN-REDD.

UN-REDD Vietnam. 2013. *Tree Allometric Equation Development for Estimation of Forest Above-Ground Biomass in Viet Nam*. Hanoi: UN-REDD Vietnam.

USAID. 2009. "Lam Dong Province Records First Hydropower Payments for Forest Conservation under Pilot Environmental Services Policy." *Vietnam.usaid.gov*. http://vietnam.usaid.gov/?q=node/59.

VACNE. 2004. *Vietnam Environment and Life*. Hanoi: Vietnam Association for Conservation, Nature and Environment.

Văn Ngọc Thành, ed. 2005. *Lịch sử lâm trường Hương Sơn: nay là công ty lâm nghiệp và dịch vụ Hương Sơn, Hà Tĩnh, 1955–2005* [History of the Huong Son state forest enterprise, 1955–2005]. Hà Tĩnh: Công ty lâm nghiệp và dịch vụ Hương Sơn [Huong Son forestry and service company].

Vandergeest, Peter, and Nancy Lee Peluso. 1995. "Territorialization and State Power in Thailand." *Theory and Society* 24 (3): 385–426.

———. 2006a. "Empires of Forestry: Professional Forestry and State Power in Southeast Asia, Part 1." *Environment and History* 12: 31–64.

———. 2006b. "Empires of Forestry: Professional Forestry and State Power in Southeast Asia, Part 2." *Environment and History* 12: 359–93.

Veilleux, Christine. 1994. "Vietnam's Forests: Historical Perspective on a Major Issue for Sustainable Development." *Journal of Business Administration* 22: 337–54.

Vietnam News Service. 1997. "PM Wants More Forest Education." *Vietnam News*, March 5, p. 3.

———. 1999. "Lone Crusader Gets His Day in Court." *Vietnam News*, April 10, p. 5.

———. 2000. "Safeguarding Viet Nam's Forests Is a Job as Dangerous as It Is Crucial." *Vietnam News*, September 9, p. 6.

VietNamNet. 2008. "Lam Dong Province to Plant New Trees for Old." *VietnamNet* Online, October 19, 2008. http://vietnamnews.vn/agriculture/181431/lam-dong-province-to-plant-new-trees-for-old.html.

———. 2014. "Who has Eliminated 130,000 Hectares of Forests in Central Highlands?" *VietnamNet* Online, March 15, 2014. http://english.vietnamnet.vn/fms/environment/97522/who-has-eliminated-130-000-hectares-of-forests-in-central-highlands-.html.

Võ Quý. 1985. "Rare Species and Protection Measures Proposed for Vietnam." In *Conserving Asia's Natural Heritage: The Planning and Management of Protected Areas in the Indomalayan Realm*, edited by J.W. Thorsell, 98–104. Geneva: International Union for the Conservation of Nature.

Von Meyenfeldt, C. F. W. M., D. Noordam, H. J. F. Savenije, E.B. Sheltens, K. van der Torren, P. A. Visser, and W. B. de Voogd. 1978. *Restoration of Devastated Inland Forests in South Vietnam.* Wageningen: Agricultural University.

VWP [Vietnam Workers Party]. 1963. *Offensive Against Poverty and Backwardness.* Hanoi: Vietnam Workers Party.

Waage, Edda, and Karl Benediktsson. 2010. "Performing Expertise: Landscape, Governmentality and Conservation Planning in Iceland." *Journal of Environmental Policy & Planning* 12 (1): 1–22.

Werner, Jane, George Dutton, and John Whitmore. 2012. *Sources of Vietnamese Tradition.* New York: Columbia University Press.

Wege, David, Adrian Long, Mai Kỳ Vinh, Vũ Văn Dũng and Jonathan Eames. 1999. *Expanding the Protected Areas Network in Vietnam for the 21st Century: An Analysis of the Current System with Recommendations for Equitable Expansion.* Hanoi: Birdlife International.

West, Paige. 2006. *Conservation is Our Government Now: The Politics of Ecology in Papua New Guinea.* Durham: Duke University Press.

Westing, Arthur. 1971. "Ecological Effects of Military Defoliation on the Forests of South Vietnam." *BioScience* 21 (17): 893–98.

———. 1974. "Postwar Forestry in North Vietnam." *Journal of Forestry* 72 (3): 153–56.

Whatmore, Sarah. 2002. *Hybrid Geographies: Natures Cultures Spaces.* Thousand Oaks, CA: Sage Publications Limited.

White, Christine. 1981. "Agrarian Reform and National Liberation in the Vietnamese Revolution: 1920–1957." PhD Thesis, Cornell University.

White, John. 1824. *A Voyage to Cochin China.* London: Longman, Hurst, Rees, Orme, Brown, and Green.

Whited, Tamara. 2000a. "Extinguishing Disaster in Alpine France: The Fate of Reforestation as Technocratic Debacle." *GeoJournal* 51 (3): 263–70.

———. 2000b. *Forests and Peasant Politics in Modern France.* New Haven: Yale University Press.

Winichakul, Thongchai. 1994. *Siam Mapped: A History of the Geo-Body of a Nation.* Honolulu: University of Hawai'i Press.

Worby, Eric. 1994. "Maps, Names, and Ethnic Games: The Epistemology and Iconography of Colonial Power in Northwestern Zimbabwe." *Journal of Southern African Studies* 20 (3): 371–92.

Wong, Teresa and Claudio Delang. 2007. "What Is a Forest? Competing Meanings and the Politics of Forest Classification in Thung Yai Naresuan Wildlife Sanctuary, Thailand." *Geoforum* 38 (4): 643–54.

Woodhouse, Philip. 2003. "African Enclosures: A Default Mode of Development." *World Development* 31 (10): 1705–20.

Woodside, Alexander. 1995. "Central Viet Nam's Trading World in the Eighteenth Century as Seen in Le Quy Don's 'Frontier Chronicles.'" In *Essays into Vietnamese Pasts*, edited by Keith Taylor and John Whitmore, 157–72. Ithaca: Cornell University Press.

World Bank. 1998. *Program 327 Review for the Vietnam Rural Development Strategy.* Hanoi: World Bank Vietnam.

World Bank. 2014. *State and Trends of Carbon Pricing.* Washington, DC: World Bank.

Wunder, Sven. 2005. *Payments for Environmental Services: Some Nuts and Bolts*. Bogor, Indonesia: Center for International Forestry Research.

Wunder, Sven, Bùi Dũng Thể, and Enrique Ibarra. 2005. *Payment is Good, Control is Better: Why Payments for Forest Environmental Services in Vietnam Have So Far Remained Incipient*. Bogor: Center for International Forestry Research.

Xuân Lạc. 2000 "Giúp các hộ làm vườn rừng [Helping families make forest gardens]." *Nông Nghiệp* [Agriculture], March 1, p. 11.

Yvon, Florence. 2008. "The Construction of Socialism in North Vietnam: Reconsidering the Domestic Grain Economy, 1954–60." *South East Asia Research* 16 (1): 43–84.

Zalasiewicz, Jan, Mark Williams, Will Steffen, and Paul Crutzen. 2010. "The New World of the Anthropocene." *Environmental Science & Technology* 44: 2228–31.

Ziegler, Alan, Thilde Bruun, Maite Guardiola-Claramonte, Thomas Giambelluca, and Deborah Lawrence. 2009a. "Environmental Consequences of the Demise in Swidden Cultivation in Montane Mainland Southeast Asia: Hydrology and Geomorphology." *Human Ecology* 37 (3): 361–73.

Ziegler, Alan, Jefferson Fox, and Jianchu Xu. 2009b. "The Rubber Juggernaut." *Science* 324 (5930): 1024–25.

Zierler, David. 2011. *The Invention of Ecocide: Agent Orange, Vietnam, and the Scientists Who Changed the Way We Think About the Environment*. Athens: University of Georgia Press.

Zimmerer, Karl. 2000. "The Reworking of Conservation Geographies: Nonequilibrium Landscapes and Nature-Society Hybrids." *Annals of the Association of American Geographers* 90 (2): 356–69.

Index

CULTURE, PLACE, AND NATURE

STUDIES IN ANTHROPOLOGY AND ENVIRONMENT

The Kuhls of Kangra: Community-Managed Irrigation in the Western Himalaya, by Mark Baker

The Earth's Blanket: Traditional Teachings for Sustainable Living, by Nancy Turner

Property and Politics in Sabah, Malaysia: Native Struggles over Land Rights, by Amity A. Doolittle

Border Landscapes: The Politics of Akha Land Use in China and Thailand, by Janet C. Sturgeon

From Enslavement to Environmentalism: Politics on a Southern African Frontier, by David McDermott Hughes

Ecological Nationalisms: Nature, Livelihood, and Identities in South Asia, edited by Gunnel Cederlöf and K. Sivaramakrishnan

Tropics and the Traveling Gaze: India, Landscape, and Science, 1800–1856, by David Arnold

Being and Place among the Tlingit, by Thomas F. Thornton

Forest Guardians, Forest Destroyers: The Politics of Environmental Knowledge in Northern Thailand, by Tim Forsyth and Andrew Walker

Nature Protests: The End of Ecology in Slovakia, by Edward Snajdr

Wild Sardinia: Indigeneity and the Global Dreamtimes of Environmentalism, by Tracey Heatherington

Tahiti Beyond the Postcard: Power, Place, and Everyday Life, by Miriam Kahn

Forests of Identity: Society, Ethnicity, and Stereotypes in the Congo River Basin, by Stephanie Rupp

Enclosed: Conservation, Cattle, and Commerce among the Q'eqchi' Maya Lowlanders, by Liza Grandia

Puer Tea: Ancient Caravans and Urban Chic, by Jinghong Zhang

Andean Waterways: Resource Politics in Highland Peru, by Mattias Borg Rasmussen

Conjuring Property: Speculation and Environmental Futures in the Brazilian Amazon, by Jeremy M. Campbell

Forests Are Gold: Trees, People, and Environmental Rule in Vietnam, by Pamela D. McElwee